Cost Records for Construction Estimating

by
W.P. Jackson

Craftsman

Craftsman Book Company
6058 Corte del Cedro Carlsbad, CA 92008

Library of Congress Cataloging in Publication Data

Jackson, W. P.
     Cost records for construction estimating.

     Includes index.
     1. Building--Estimates. 2. Business records.
I. Title.
TH437.J33 1984   692'.5   84-14271
ISBN 0-910460-41-8

Contents

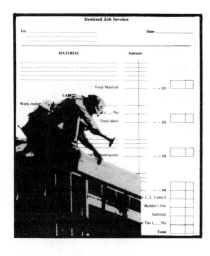

Chapter 1

Keeping Accurate Cost Records

Do you know of any construction companies that have gone broke?

If you've been around the construction industry very long, you've heard of contractors going belly up . . . leaving jobs unfinished and suppliers unpaid. Every time interest rates climb a few points, construction activity falls off, or shortages develop in key materials. Many general contractors and subcontractors have to struggle to meet their payroll and stay in business.

To the contractor who's short of cash, it seems that the problem is either a lack of ready capital or a burdensome debt load. But most consultants that have studied failures in the construction industry agree that it isn't cash shortages that shut down most contractors. Working capital requirements are relative. There's some level of business that's right for the capital that's available, no matter how meager it may be.

And the problem isn't a lack of construction work or poor construction skills. Some very busy builders with first-rate knowledge of their trade have gone under, even at times when other contractors were making good money.

There's one characteristic that's common to most failures in the construction industry: poor record-keeping. When a builder is teetering on the edge of bankruptcy, the consultant called in to bail him out will nearly always find chaotic records. If there are any records at all!

The contractor gets into a financial bind because he doesn't know what his costs are. His estimates are based more on guesswork than on known costs. He's probably had trouble with the I.R.S. because payroll records were incomplete and tax reports and deposits were not made on time. Often there will be no complete file of payables and receivables. Material receiving records will be missing or inadequate.

If what I've just described fits your business, you're a good candidate for the system I'm going to describe in this book. Every construction contractor needs some simple and efficient system for controlling and recording project costs. If your cost records aren't providing all the information they should, and if you're worried that the documents in your file don't meet I.R.S. standards, keep reading. This book's for you.

Your cost records should be the foundation for every construction estimate you compile. Sure, many builders use cost-estimating manuals. But nobody will ever sell a cost book that comes close to matching the accuracy of your own records of work completed on your jobs, by your own crews, and under your supervision.

But, you say, "Don't I need an accountant to keep all those numbers?"

Fortunately, you don't need an accountant or bookkeeper to record construction costs. If you can keep time for your employees, you can keep accurate cost records and with a minimum of effort and expense.

Why Keep Good Cost Records?

Keeping accurate cost records will help put extra profit into your pocket. Here's how. Accurate cost records will:

1) Keep you up-to-date on the expenses for each job all during construction. You can't control costs if you don't know what the costs are. Good records show all your costs and make control possible.

2) Tell you immediately where costs are exceeding estimates.

3) Provide the essential data for estimating future jobs.

4) Help you supervise and schedule construction. (Cost records show the manhours required for each phase of construction.)

5) Reveal waste in materials and manhours: excess crew members, inefficiency, unnecessary waiting for materials, time lost due to weather problems, and other job factors that may be beyond your control.

The beginning of every good cost-record system is an organized account system—usually a set of numbered categories into which all job costs fall. The format followed by most professional construction companies is The Uniform Construction Index. It's published jointly by the American Institute of Architects, the Associated General Contractors and the Construction Specifications Institute.

The U.C.I. lays out an extensive cost analysis format and system of account numbers for every part of a construction project. For example, U.C.I. account number 06190 is used for recording all costs associated with wood trusses. This format is essential for contractors who handle major construction projects. But most residential builders don't need that much detail. You can get the same results with a simplified account number system that you design yourself.

The format you use to keep cost records isn't critical. *But using one system consistently is vital to your success.* The basic objective is to know your costs at every stage of the job and with a minimum of record-keeping.

In this chapter, we'll look briefly at automated record-keeping. Then we'll explain each of the forms you'll need for an efficient cost record system.

Automated Record-keeping

The inexpensive personal computer is revolutionizing the way this country does business. And the construction industry is no exception. Every builder dreams of the day when a computer will maintain all of his records automatically, at high speed and for peanuts. We may see that day. But it isn't here yet for most small construction companies.

Big construction companies have been very successful in putting most record-keeping on computer. But they have the resources to develop custom programs that can be used by employees who do nothing else but run the computer.

Most smaller contractors can't and shouldn't commit thousands of dollars and hundreds of hours to automating their record-keeping systems. They don't have a full-time data processing staff to keep the equipment and programs running. And they can't afford the luxury of developing the custom programs required to fit their needs.

Instead, you'll have to select from among standard programs that reflect someone else's concept of how a construction business should be run. Unfortunately, that always involves compromise. And you don't find out if the standard programs really meet your needs until after you've invested the time and money to have the programs installed.

If you enjoy using computers and have one at your disposal, consider using it to do some basic accounting. That's an easy application. Many good programs are available. Some will fit your business very well. If you feel comfortable with the accounting programs, next consider a payroll program. Eventually, you may even be able to buy a job cost program that does most of what we describe in this book.

But don't expect any computer to automate your record-keeping overnight. Even if you have the perfect programs for every application you need, you'll still have to gather the records and organize them just the way we describe in this book. The computer won't replace record-keeping. It'll just report the information faster.

It's good practice to use a computer for payroll and general ledger (bookkeeping). But that doesn't mean you have to buy a computer if you don't already own one. Many banks and service bureaus offer computerized payroll systems for about $2 per employee per pay period. Some banks even waive this fee if you maintain a specified average balance in your checking account.

Here's how these payroll systems work. You supply the bank or service bureau with the name, pay rate and social security number for each employee. A day or two before payday, you phone in the number of hours each employee worked. On payday, the checks are available for pickup, or they can be delivered by messenger. All deductions are computed, tax deposits are made, and forms are filed with state and federal authorities. This simplifies your payroll chores and gives you time to concentrate on more productive work.

A service bureau (or bookkeeping service) will also make up a monthly balance sheet and income statement. You supply a copy of all checks written and all deposits made during the month. Itemize charges made to important accounts. About 10 days after submitting this information, you'll receive a report that summarizes operations for the month and for year-to-date. The cost will be about $50 per month.

Remember that a service bureau can reduce your paperwork, but it's still your job to gather and organize the information the bureau will use to prepare the payroll and general ledger reports. Payroll will be covered in more detail in Chapter 2.

We've discussed the reasons for keeping accurate cost records. And we've looked briefly at automated record-keeping. Now let's look at each of the forms you'll need to set up your own cost record-keeping system.

Forms for Simplified Record-keeping

Every contractor needs some good forms for keeping his records. You can buy most of them at any office supply store. Two forms, however, should be custom designed: the weekly time sheet with a daily log, and a payroll envelope with a payroll receipt. Two additional forms you'll need are: the individual employee earnings record, and columnar (or analysis) sheets. Let's look at each of these forms in detail.

Blank copies of the customized forms used here are provided in the back of this manual. Make copies for your own use on your office copier, or have a print shop make you a pad of them.

Weekly Time Sheet

The weekly time sheet is your payroll sheet. It lists employees who are entitled to compensation. It tells how much each employee should receive for the pay period.

It's easy to make errors in recording employee time. Make sure that each employee is given credit for the correct number of hours worked each day.

At the bottom of each weekly time sheet, there's room to describe the work done each day. Record this information at the end of each day while it's fresh in your mind. This daily log is the most essential cost record and becomes the source document for many other cost records.

Keep separate weekly time sheets for each job. If one of your employees works part of a day on one job and part on another, you must record his hours on two different time sheets. Here's how.

Figures 1-1A and 1-1B show employee time charged to two different jobs. Of the six employees listed during this pay period, four of them worked on both jobs.

Look at Figure 1-1A. On Monday, three men each worked 8 hours (24 manhours) doing the formwork for the footings. On Tuesday, they worked another 24 manhours. On Wednesday, the formwork was finished after only 18 manhours. So on this job, it took 66 manhours to construct the forms for the footings.

This is your record of the actual labor cost for the formwork. Compare it to your estimated labor cost for the formwork. Use this record as a reliable guide when making future estimates. But make sure you use only the number of manhours (and not the dollars) from this record when you prepare future labor estimates. Why? Because taxes, labor rates, and insurance rates change constantly, whereas manhours have some degree of stability.

Labor costs are always estimates. But labor productivity, while it does vary with each journeyman, and from day to day — up one day and down the next, depending on the worker's attitude and how he feels that day — can be fairly accurately estimated by an experienced estimator who can anticipate potential problems and opportunities in each estimate.

Payroll Envelope and Receipt

You must keep a record of each employee's gross earnings, deductions and net pay for each pay period throughout the year. As an employer, you'll need these figures for your cost records and for filing your quarterly and year-end tax returns. Your employee will need these figures for his personal records and for filing his own tax returns.

Figure 1-2A shows a payroll envelope and receipt. The employee keeps the envelope, which shows gross earnings, deductions and net pay. The receipt, which provides the same information, is kept by the employer. The employee signs the receipt if he is paid in cash. If he's paid by check, he doesn't need to sign the receipt, because the cancelled check is a receipt.

Page____of____pages

Weekly Time Sheet

For period ending _8-28-XX_   _BROWN_____job

	Name	Exemptions	Days 23 M	24 T	25 W	26 T	27 F	28 S	Rate	Hours worked Reg.	Over-time	Total earnings
1	D. White		8	8	6	6½	2½	X		31		
2	J. Kidd		8	8	6	6½	2½	X		31		
3	R. Farlow		8	8	6	6½	X	X		28½		
4	N. Neel		X	X	X	6½	X	X		6½		
5												
6												
7												
8												
9												
10												
11												
12												
13												
14												
15												
16												
17												
18												
19												
20												

Daily Log

Monday FORMWORK FOR FOOTINGS

Tuesday FORMWORK FOR FOOTINGS

Wednesday FINISHED FORMS (READY TO POUR CONCRETE)

Thursday POURED CONCRETE FOR FOOTINGS

Friday REMOVED FORMS FROM FOOTINGS (RAIN)

Saturday X X X

Weekly Time Sheet - Brown Job
Figure 1-1A

Weekly Time Sheet

Page____of____pages

For period ending __8-28-XX__   __GREEN__ job

	Name	Exemptions	23 M	24 T	25 W	26 T	27 F	28 S	Rate	Reg.	Over-time	Total earnings
			Days							**Hours worked**		
1	D. White		X	X	2	1½	5½	X		9		
2	J. Kidd		X	X	2	1½	5½	X		9		
3	R. Farlow		X	X	2	1½	8	X		11½		
4	N. Neel		8	8	8	1½	8	X		33½		
5	C. Lawson		8	8	8	8	8	X		40		
6	R. Leedy		8	8	8	8	8	X		40		
7												
8												
9												
10												
11												
12												
13												
14												
15												
16												
17												
18												
19												
20												

Daily Log

Monday __Inside trim__

Tuesday __Inside trim__

Wednesday __Inside trim__

Thursday __Inside trim__

Friday __Inside trim__

Saturday __X X X__

Weekly Time Sheet - Green Job
Figure 1-1B

Envelope side:

Date _____

Employee _____

Number of withholding
exemptions _____

____Hours on _____job

____Hours on _____job

____Hours on _____job

____Total hours @_____ = $ _____

Deductions:

Social Security $_____

Federal withholding tax $_____

State withholding tax $_____

Other charges $_____

    Total Deductions $_____

    Net Amount $_____

Employee
Signature _____

Envelope

Receipt side:

Date _____

Employee _____

Number of withholding
exemptions _____

____Hours on _____job

____Hours on _____job

____Hours on _____job

____Total hours @_____ = $ _____

Deductions:

Social Security $_____

Federal withholding tax $_____

State withholding tax $_____

Other charges $_____

    Total Deductions $_____

    Net Amount $_____

Employee
Signature _____

Receipt

**Payroll Envelope and Receipt
Figure 1-2A**

Make sure that your payroll envelope and receipt provide the following information:

1) Date (This will be the date at the end of the pay period.)

2) Name of employee

3) Number of withholding exemptions

4) Number of hours worked on each job during the pay period

5) Calculation showing: total hours worked, times the rate of pay, equals the employee's gross pay

6) Deductions: social security (FICA), federal withholding tax, state withholding tax, other deductions and total deductions

7) Net amount: total deductions subtracted from the employee's gross pay. This is the amount of money the employee actually receives for this pay period

8) Employee's signature

Figure 1-2B shows a filled-in receipt detached from an employee's payroll envelope. It's especially important that every paid-in-cash employee sign the receipt, and that you keep these receipts. As an employer, you can be challenged months (or even years) later about underpayment on a particular job. The receipts can help resolve future disputes.

Date ___8-28-XX___

Employee ___D. L. White___

Number of withholding
exemptions ___1___

_31_ Hours on ___Brown___ job

_9_ Hours on ___Green___ job

___ Hours on _____ job

_40_ Total hours @ _$25.00_ = $ ___$1,000.00___

Deductions:

Social Security $ _67.00_

Federal withholding tax $_251.80_

State withholding tax $ _51.75_

Other charges $ _-0-_

    Total Deductions $ _370.55_

    Net Amount $ _629.45_

Employee
Signature ___D. L. White___

**Payroll Receipt
Figure 1-2B**

Individual Employee Earnings Record

The information shown on the payroll receipt is entered on the individual employee earnings record shown in Figure 1-3. The information on this record will be required for your quarterly and year-end reports to the state and federal governments. Quarterly reports are filed at the end of each quarter. Let's take a look at these important federal and state reports.

The Employer's Quarterly Federal Tax Return (Form 941) requires the employee's name, social security number, and taxable FICA wages (before deductions), plus the total wages paid to all employees for the quarter.

You must also record the FICA tax and the total amount of income tax withheld from all employees for the quarter. The FICA tax includes both tax paid by the employer and tax withheld from each employee's earnings. You'll have to pay these taxes at the same time you file the report.

The Internal Revenue Service publishes Circular E, Employer's Tax Guide. It explains withholding procedures, and outlines the required filing dates. Your accountant will also help you meet federal employer tax requirements.

You must also file two quarterly reports with your state government. One is a report showing the total amount of state income tax withheld from your employees for the quarter. This tax must be paid at the time the report is filed.

The second report is for the FUTA tax (Federal Unemployment Tax Act). The report is filed with your state employment commission. This report requires the employee's name, social security number, and total wages for the quarter. Multiply the quarterly wages times the tax rate, and that's the amount you'll have to pay.

In addition to the quarterly reports, you must file year-end reports with the federal and state governments. These reports include the Wage and Tax Statements (Form W-2) and the Employer's Annual Federal Unemployment Tax Return (Form 940). They also include a summary of state income taxes withheld for all employees for the year.

You can see how important it is to keep a record of the quarterly earnings and deductions for each of your employees. These records are indispensable. Sifting through payroll receipts at the end of each quarter is neither cost-effective nor professional. Let the individual employee earnings record do the work for you.

Look at Figure 1-3. This simplified form will give you all the information necessary for the quarterly and year-end reports requiring your employees' earnings and deductions.

The form is divided into four quarters. Each line shows the number of hours worked for each pay period, total earnings for the pay period, itemized deductions, and net pay. All of this information is taken from the payroll receipt shown in Figure 1-2B.

The payroll receipt shows a total of 40 hours worked for the week ending August 28. This is the thirty-fifth week in the year. The date 8/28 is shown after week number 35 in the third quarter.

In the next column, enter the employee's total number of hours for the week. In our example, D. L. White worked 40 hours. He didn't work any overtime. If he had worked overtime, his regular time plus his overtime would be shown here.

In the total earnings column, enter the figure $1,000.00 from the payroll receipt. The deductions are $67.00 for social security, $251.80 for federal

NAME OF EMPLOYEE	*D.L. White*			SOCIAL SECURITY NUMBER	*000-00-000*
ADDRESS				CITY OR TOWN	
DATE OF BIRTH		MARRIED ☒ OR SINGLE ☐	NUMBER OF EXEMPTIONS	PHONE NO.	CLOCK NO.
POSITION *Carpenter*	RATE *$25.00*	DATE		DATE STARTED *3-27-XX*	DATE TERMINATED
REMARKS				REASON	

FIRST QUARTER 19XX / SECOND QUARTER 19XX

WEEK #	DATE	HOURS WORKED REG.	OVER-TIME	TOTAL EARNINGS	SOC. SEC.	FED. WITH. TAX	STATE WITH. TAX			NET PAY	WEEK #	DATE	HOURS WORKED REG.	OVER-TIME	TOTAL EARNINGS	SOC. SEC.	FED. WITH. TAX	STATE WITH. TAX			NET PAY
1											14										
2											15										
3											16										
4											17										
5											18										
6											19										
7											20										
8											21										
9											22										
10											23										
11											24										
12											25										
13											26										
Total 1st Qtr											Total 2nd Qtr										
TOTAL 3 MOS											TOTAL 6 MOS										

THIRD QUARTER 19XX / FOURTH QUARTER 19XX

WEEK #	DATE	HOURS WORKED REG.	OVER-TIME	TOTAL EARNINGS	SOC. SEC.	FED. WITH. TAX	STATE WITH. TAX			NET PAY	WEEK #	DATE	HOURS WORKED REG.	OVER-TIME	TOTAL EARNINGS	SOC. SEC.	FED. WITH. TAX	STATE WITH. TAX			NET PAY
27											40										
28											41										
29											42										
30											43										
31											44										
32											45										
33											46										
34											47										
35	8/28	40		1,000.00	67.00	251.80	51.75			629.45	48										
36											49										
37											50										
38											51										
39											52										
Total 3rd Qtr											Total 4th Qtr										
TOTAL 9 MOS											TOTAL YEAR										

Individual Employee Earnings Record
Figure 1-3

withholding tax, and $51.75 for state withholding tax. The net amount shown on the payroll receipt is entered in the net pay column on the individual employee earnings record.

This information is entered every week (or pay period). At the end of each quarter, the employee's earnings, social security deductions, federal and state withholding tax deductions, and other deductions are totaled. At the end of the year, all of the quarterly figures are totaled for the year-end reports.

Columnar or Analysis Sheets
You can purchase blank columnar or analysis sheets at any office supply store. The number of jobs will determine the number of columns and the size of sheet required. The 10-column sheet shown in Figure 1-4 can be purchased in 8½'' x 14'' pads.

Use these analysis sheets for keeping three important records:

- Income and disbursements
- Consolidated report of labor costs per job
- Subcontracted work

Income and disbursements— Here's how to use the analysis sheet shown in Figure 1-4. Record the date of each entry in the column at the left. The column headed receipts and disbursements is for the name of the person or company that money was disbursed to or received from. The check number appears in the next column.

The income column (column 1) is used only for money received. The disbursements column (column 2) is for the total amount of money paid out. Column 2 (on each line) is equal to the total of columns 3 through 8 (on each line).

If a disbursement is for more than one job, it's distributed to each job in its proper column. For example, let's say we bought material from one lumber company for two different jobs during the month. The material for one job amounted to $675.85. The material for the other job amounted to $1,575.50. The total amount we owe is $2,251.35.

Now let's pay our bill. First, we write a check for $2,251.35. On our analysis sheet, we enter $2,251.35 in column 2. In the proper column for one job we enter $675.85, and in the proper column for the other job we enter $1,575.50.

As shown in Figure 1-4, you should keep running totals (in pencil) for columns 1 through 8. If the running total for column 2 is not equal to the total of the running totals from columns 3 through 8, then an error has been made somewhere in the entries or additions and must be corrected at once. By keeping the running totals up-to-date, you always know what your costs are.

Consolidated report of labor costs per job— You should know your total labor costs on every job at any given time. The consolidated report of labor costs per job will give you this information. See Figure 1-5.

The date shown in Figure 1-5 is the date the payroll period ended. It must be the same date as shown on the payroll receipt and the individual employee earnings record.

Use column 1 to record the gross wages for the pay period. Columns 2, 3 and 4 represent the portion of gross wages charged to each job. The total of columns 2, 3 and 4 (on each line) must equal column 1 (on each line). Keep running totals (in pencil) so that you'll have a complete picture of your labor costs at all times.

In Figure 1-5, the total gross wages (column 1) for the payroll period ending March 12 were $5,487.50. This amount was charged as follows:

Baker job (column 2)	$3,087.50
Brown job (column 3)	2,362.50
Green job (column 4)	37.50
Total	$5,487.50

Since the total of columns 2, 3 and 4 equals the total in column 1, we can be sure that no errors have been made in the entries or additions. Another check for errors is: The total gross wages must always equal the total earnings for all employees as shown on the individual employee earnings record (Figure 1-3).

This cross-check is helpful when you're paying the insurance premiums on worker's compensation and liability insurance for your employees. It's also a safeguard against errors when you file your quarterly and year-end reports.

Subcontracted work— Figure 1-6 shows a record of the money paid for subcontracted work on each job.

In Figure 1-6, the total disbursements (column 1) through June 10 were $4,447.70. This amount was charged to the following jobs:

Baker job (column 2)	$ 753.80
Brown job (column 3)	1,634.70
Green job (column 4)	2,059.20
Total	$4,447.70

	Date		Receipts and Disbursements	CK. No.	1 Income	2 Disbursements	3 Interest
1	—		Balance Forward	—	175026 11	33269 70	
2	5-	28	J.W. Green	6104	4369 60		
3	6-	3	Bailey Lumber Co.	2086		3788 75	
4	6-	4	Southern Insurance Co.	2085		1329 00	
5	6-	4	Exxon	Cash		12 50	
6	6-	6	IBM Corporation	2096		56 16	
7					179395 71	38456 11	
8							
9							
10							
11							
12							
13							
14							
15							
16							
17							
18							
19							
20							
21							
22							
23							
24							
25							
26							
27							
28							
29							
30							

Income and Disbursements

4	5	6	7	8 Green Job	9 10 Explamation
Taxes	Office	Vehicle	Insurance		
6855 83	590 02	997 52	1660 00	2316 6 33	
				3788 75	May Invoices
			1329 00		Addition Premium Charge to Green Job
		12 50			13.5 Gals. Gasoline
	56 16				Typewriter Supplies
6955 83	646 18	1010 02	2989 00	26955 08	

Figure 1-4

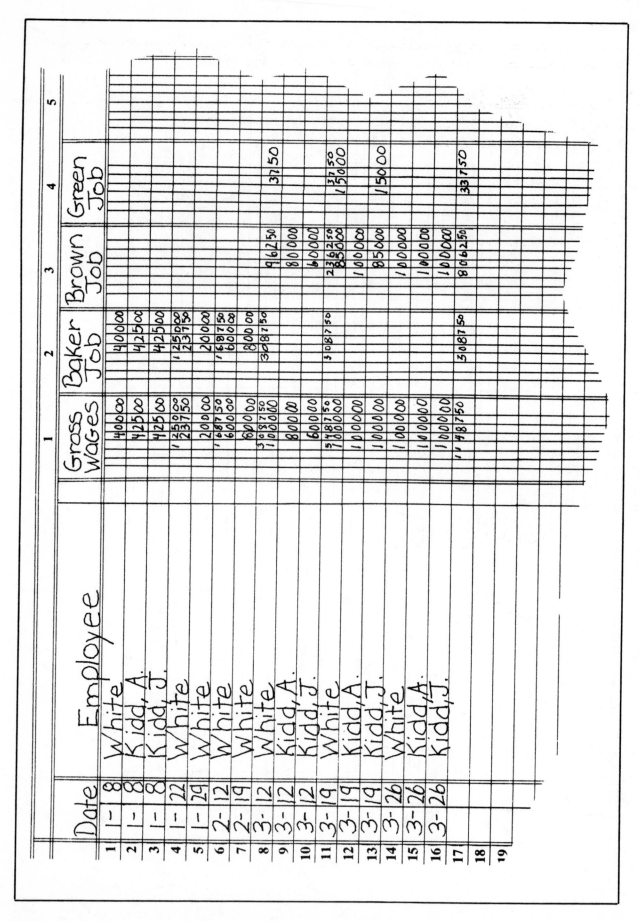

Consolidated Report of Labor Costs per Job
Figure 1-5

	Date	Subcontractor	Ck No	Disbursement	Baker Job	Brown Job	Green Job
				1	2	3	4
1	2-25	D + A Plumbing	1816	358 00	358 00		
2	3-24	H. + H. Drywall Co.	1840	748 80		748 80	
3	5-13	C. W. Martin	1893	395 80	395 80		
4	5-13	R. W. Waldron Co	1896	292 20		292 20	
5	5-30	D. W. Via	1911	274 05		274 05	
6	5-30	J. C. Short	1912	65 25		65 25	
7	6-2	C. W. Martin	1916	2059 20			2059 20
8	6-10	H. + H. Drywall Co	1925	254 40		254 40	
9				4447 70	753 80	1634 70	2059 20
10							

Subcontract Work
Figure 1-6

When you need a cost breakdown on the subcontracted work for each phase of construction, just refer to your subcontract work record.

Look again at Figure 1-6. Let's do a sample calculation for the subcontracted work on the Brown job. We'll look at the phase of construction from March 24 through June 10. Here's how it looks:

Drywall (line 2, dated 3/24)	$ 748.80
Concrete (line 4, dated 5/13)	292.20
Stonework (line 5, dated 5/30)	274.05
Stonework (line 6, dated 5/30)	65.25
Drywall (line 8, dated 6/10)	254.40
Total	$1,634.70

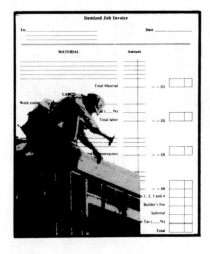

Chapter 2

Payroll Records

Now let's apply what we've learned. In this chapter, we'll create a simple, efficient payroll record system using the four payroll forms introduced in Chapter 1. Then we'll look at each of the important quarterly and year-end reports. Your new payroll record system will simplify the preparation of these reports. We'll show you how.

Efficient Payroll Record System
The four important payroll forms are: weekly time sheet; payroll envelope and receipt; individual employee earnings record; and consolidated report of labor costs per job. Now let's use these four forms to create an efficient payroll record system.

Weekly Time Sheet
Keeping payroll records begins with the weekly time sheet. As we discussed in Chapter 1, you'll need separate time sheets for each job. This is your way of determining the labor costs on a per-job basis.

Fill out the daily log at the bottom of each sheet every day. It's important to do this while the information is fresh in your mind. This record provides you with a complete picture of your actual costs. If your costs are getting out of line, you'll know it immediately.

Figure 2-1A shows the weekly time sheet and daily log for the Baker job for the pay period ending January 8. This job is nearly finished, but winter weather has interrupted construction. On Monday, only one hour was worked by each employee, and this was cleanup work. On Tuesday, no work was done because the weather was too severe. This information is recorded in the daily log. On Wednesday and Thursday, the weather was moderate enough for three workmen to install siding on the front porch. On Friday, it was too cold for outside work.

Figures 2-1B through 2-1K show additional examples of weekly time sheets. These time sheets span a period of one year and cover three different jobs: the Baker, Brown and Green jobs. Each job is at a different stage of construction: the Baker job is nearly completed; the Brown job is about half finished; and the Green job is just beginning. As the year progresses, new employees are added to the weekly time sheets and to all other payroll records.

Weekly Time Sheet

Page____of____pages

For period ending __1-8-xx__   BAKER __job

	Name	Exemptions	Days JANUARY 3 M	4 T	5 W	6 T	7 F	8 S	Rate	Reg.	Over-time	Total earnings	
1	A.L. KING		1	X	8	8	X	X	25.00	17		425	00
2	J.E. KING		1	X	8	8	X	X	25.00	17		425	00
3	D.L. WHITE		X	X	8	8	X	X	30.00	16		480	00
4										TOTAL		1330	00
5													
6													
7													
8													
9													
10													
11													
12													
13													
14													
15													
16													
17													
18													
19													
20													

Daily Log

Monday __COLD --- CLEANED OUT GARAGE__
Tuesday __COLD --- SNOW__
Wednesday __SIDING, TRIM ON FRONT PORCH__
Thursday __SIDING, TRIM ON FRONT PORCH__
Friday __COLD --- SNOW__
Saturday __X X X X__

Weekly time sheet
Figure 2-1A

Weekly Time Sheet

Page____of____pages

For period ending __3-12-XX__   __BROWN__ job

Name	Exemptions	Days MARCH 7 M	8 T	9 W	10 T	11 F	12 S	Rate	Hours worked Reg.	Over-time	Total earnings	
1 D.L. WHITE		8	8	6½	8	8	X	30.00	38 ½		1155	00
2 A.L. KING		X	8	8	8	8	X	25.00	32		800	00
3 J.E. KING		X	X	8	8	8	X	25.00	24		600	00
4									TOTAL--$2555			00
5												
6												
7												
8												
9												
10												
11												
12												
13												
14												
15												
16												
17												
18												
19												
20												

Daily Log

Monday __WATER PROOFING ON FOUNDATION__
Tuesday __WATER PROOFING ON FOUNDATION ··· FILL DIRT IMPORTED__
Wednesday __DRAIN TILE AROUND FOUNDATION ··· FILL DIRT IMPORTED__
Thursday __O/S TRIM, COLUMN POSTS ON FRONT PORCH, DRY WALL FINISH__
Friday __SIDING, O/S TRIM ··· FILL DIRT IMPORTED__
Saturday __— — — — — — —__

Weekly time sheet
Figure 2-1B

Weekly Time Sheet

Page____of____pages

For period ending 3-12=XX

____GREEN____ job

	Name	Exemptions	Days MARCH						Rate	Hours worked		Total earnings	
			7 M	8 T	9 W	10 T	11 F	12 S		Reg.	Over-time		
1	D.L. WHITE		X	X	1½	X	X	X	30.00	1½		45	00
2													
3													
4													
5													
6													
7													
8													
9													
10													
11													
12													
13													
14													
15													
16													
17													
18													
19													
20													

Daily Log

Monday ___ X X X X ___
Tuesday ___ X   X   X X ___
Wednesday ___TOOK LOT ELEVATIONS FOR PREPARING PLOT PLAN___
Thursday___ X   X   X X ___
Friday ___ X X   X   X ___
Saturday ___ X X X X ___

Weekly time sheet
Figure 2-1C

Weekly Time Sheet

Page____of____pages

For period ending ___4 - 16 - XX___

___GREEN___ job

	Name	Exemptions	Days APRIL						Rate	Hours worked		Total earnings	
			11 M	12 T	13 W	14 T	15 F	16 S		Reg.	Over-time		
1	D.L. WHITE		8	8	8	8	8	—	30.00	40	—	1200	00
2	J.E. KING		3	X	X	X	X	—	25.00	3	—	75	00
3	D.L. WEST		X	8	8	8	8	—	25.00	32	—	800	00
4											TOTAL --	2075	00
5													
6													
7													
8													
9													
10													
11													
12													
13													
14													
15													
16													
17													
18													
19													
20													

Daily Log

Monday LAYED OUT FOUNDATION (EXCAVATION BY SUBCONTRACTOR)
Tuesday FOOTING (WATER LINES BY SUB CONTRACTOR)
Wednesday FOOTING (WATER LINE FROM METER TO HOUSE INSTALLED)
Thursday FOOTING (WATER LINE COVERED)
Friday FOOTING (GRAVEL FOR TEMPORARY DRIVEWAY EMPLACED)
Saturday X X X X

Weekly time sheet
Figure 2-1D

Page____of____pages

Weekly Time Sheet

For period ending ___4-16-XX___   ___BROWN___job

| | Name | Exemptions | Days APRIL | | | | | | Rate | Hours worked | | Total earnings | |
			11 M	12 T	13 W	14 T	15 F	16 S		Reg.	Over-time		
1	J.E. KING		5	8	8	8	8	X	25.00	37	—	925	00
2													
3													
4													
5													
6													
7													
8													
9													
10													
11													
12													
13													
14													
15													
16													
17													
18													
19													
20													

Daily Log

Monday ___O/S TRIM (FILL DIRT IMPORTED BY SUBCONTRACTOR)___

Tuesday ___O/S TRIM (DOZER WORK)___

Wednesday ___O/S TRIM (I/S PAINTING BY SUBCONTRACTOR)___

Thursday ___O/S TRIM (O/S PAINTING BY SUBCONTRACTOR)___

Friday ___FORMS FOR CONCRETE IN GARAGE___

Saturday ___X X X X___

Weekly time sheet
Figure 2-1E

Weekly Time Sheet

For period ending **6-25-XX**   **BROWN** job

	Name	Exemptions	Days JUNE						Rate	Hours worked		Total earnings	
			20 M	21 T	22 W	23 T	24 F	25 S		Reg.	Over-time		
1	D.L. WHITE		6½	1½	X	X	X	X	30.00	8		240	00
2	A.L. KING		6½	X	X	X	X	X	25.00	6½		162	50
3	J.E. KING		6½	1½	X	X	X	X	25.00	8		200	00
4										TOTAL--		602	50
5													
6													
7													
8													
9													
10													
11													
12													
13													
14													
15													
16													
17													
18													
19													
20													

Daily Log

Monday POURED CONCRETE PATIO AND BACK WALK
Tuesday STRIPPED FORMS AND DRESSED UP CONCRETE
Wednesday X X X X
Thursday X X X X
Friday X X X X
Saturday X X X X

Weekly time sheet
Figure 2-1F

Weekly Time Sheet

Page____of____pages

For period ending 6-25-XX   GREEN job

	Name	Exemptions	20 M	21 T	22 W	23 T	24 F	25 S	Rate	Reg.	Over-time	Total earnings	
			Days **JUNE**							**Hours worked**		**Total earnings**	
1	D.L. WHITE		X	X	7½	3	9	X	30.00	19½		585	00
2	C.A. LESTER		X	X	7½	3	9	X	30.00	19½		585	00
3	A.L. KING		X	X	7½	3	9	X	25.00	19½		487	50
4	J.E. KING		X	X	7½	3	9	X	25.00	19½		487	50
5	D.L. WEST		X	X	7½	3	9	X	25.00	19½		487	50
6									TOTAL --			2632	50
7													
8													
9													
10													
11													
12													
13													
14													
15													
16													
17													
18													
19													
20													

Daily Log

Monday FACTORY BUILT HOUSE SCHEDULED TO ARRIVE WEDNESDAY
Tuesday   X X X X
Wednesday SET PARTITIONS ON FIRST FLOOR -- RAIN
Thursday UNLOADED SECOND TRUCK -- RAIN
Friday UNLOADED THIRD TRUCK -- SET PARTITIONS ON 1st FLOOR
Saturday   X X X X

Weekly time sheet
Figure 2-1G

Weekly Time Sheet

Page____of____pages

For period ending **8-6-XX**   **GREEN** job

	Name	Exemptions	Days AUGUST 1 M	2 T	3 W	4 T	5 F	6 S	Rate	Hours worked Reg.	Over-time	Total earnings
1	R.R. LEWIS		8	8	8	8	8	X	30.00	40		1200 00
2	C.4. LESTER		8	8	8	8	8	X	30.00	40		1200 00
3	A.L. KING		8	8	8	8	8	X	25.00	40		1000 00
4	J.E. KING		8	8	8	8	8	X	25.00	40		1000 00
5	L.H. KIDD		8	8	8	8	8	X	25.00	40		1000 00
6	D.L. WEST		8	8	8	4	8	X	25.00	36		900 00
7										TOTAL - -		6300 00
8												
9												
10												
11												
12												
13												
14												
15												
16												
17												
18												
19												
20												

Daily Log

Monday WORK ON ROOF --- PLUMBINE (SUBCONTRACTOR)
Tuesday PREPARATION WORK FOR ROOFING AND BRICK MASONS
Wednesday PREPARATION WORK FOR ROOFING AND BRICK MASONS
Thursday PREPARATION WORK FOR ROOFINE AND BRICK MASONS
Friday PREPARATION WORK FOR ROOFINE AND BRICK MASONS
Saturday XX XX

Weekly time sheet
Figure 2-1H

Weekly Time Sheet

Page____of____pages

For period ending __9-24-XX__   __GREEN__ job

	Name	Exemptions	Days SEPTEMBER 19 M	20 T	21 W	22 T	23 F	24 S	Rate	Hours worked Reg.	Over-time	Total earnings	
1	C.A. LESTER		8	8	X	8	8	X	30.00	32		960	00
2	A.L. KING		8	8	X	X	X	X	25.00	16		400	00
3	J.E. KING		8	8	X	X	X	X	25.00	16		400	00
4										TOTAL –		1760	00
5													
6													
7													
8													
9													
10													
11													
12													
13													
14													
15													
16													
17													
18													
19													
20													

Daily Log

Monday __SIDING --- INSULATION - BRICK MASONS STARTED BRICK WORK__

Tuesday __BRICK WORK --- CUT OPENING FOR CHIMNEY__

Wednesday __BRICK WORK__

Thursday __WORK ON CHIMNEY -- COORDINATING WORK BY CARPENTER__

Friday __BRICK WORK --- COORDINATING WORK BY CARPENTER__

Saturday __X X X X__

Weekly time sheet
Figure 2-1I

Weekly Time Sheet

Page____of____pages

For period ending **12·24-XX**   **GREEN** job

	Name	Exemptions	December 19 M	20 T	21 W	22 T	23 F	24 S	Rate	Hours worked Reg.	Over-time	Total earnings	
1	D.L. WHITE		8	8	8	8	8	X	30.00	40		1200	00
2	A.L. KING		8	8	8	8	8	X	25.00	40		1000	00
3	J.E. KING		8	8	8	8	8	X	25.00	40		1000	00
4									TOTAL –			3200	00
5													
6													
7													
8													
9													
10													
11													
12													
13													
14													
15													
16													
17													
18													
19													
20													

Daily Log

Monday O/S AND I/S TRIM --- BRICK WORK
Tuesday I/S TRIM --- RAIN
Wednesday I/S TRIM --- COLD AND SNOW FLURRIES
Thursday I/S TRIM --- COLD
Friday EMPLOYEES GIVEN DAY OFF WITH PAY (CHRISTMAS)
Saturday X X X X

Weekly time sheet
Figure 2-1J

Weekly Time Sheet

For period ending __12-31-XY__   GREEN ____job

	Name	Exemptions	Days DECEMBER 26 M	27 T	28 W	29 T	30 F	31 S	Rate	Hours worked Reg.	Over-time	Total earnings	
1	A. L. KING		X	8	8	X	X	X	25.00	16		400	00
2	J. E. KING		X	8	8	8	8	X	25.00	32		800	00
3									TOTAL			1 200	00
4													
5													
6													
7													
8													
9													
10													
11													
12													
13													
14													
15													
16													
17													
18													
19													
20													

Daily Log

Monday __XXXX --- HOLIDAY (DAY AFTER CHRISTMAS)__
Tuesday __1/S TRIM__
Wednesday __1/S TRIM --- FINISHED CIRCULAR STAIRS__
Thursday __1/S TRIM__
Friday __INSTALLED KITCHEN CABINETS__
Saturday __XXXX__

Weekly time sheet
Figure 2-1K

Payroll Envelope and Receipt

The information recorded on the weekly time sheet is then transferred to the payroll envelope and receipt. Figure 2-2A shows three payroll receipts for the pay period ending January 8. The payroll envelope and receipt must show the following information:

1) The date on which the pay period ends. This date must be the same as the date shown on the weekly time sheet.

2) Name of employee.

3) Number of withholding exemptions.

4) Number of hours worked on each job during the pay period.

5) Total number of hours worked on all jobs during the pay period.

6) Gross pay. To calculate the employee's gross pay, multiply total hours times the rate of pay.

7) Social security and federal withholding tax deductions. (See Circular E, Employer's Tax Guide, issued by the Internal Revenue Service.) For state withholding tax deductions, consult the department that handles payroll taxes in your state.

8) Net amount of money, after all deductions, that the employee receives for the pay period. This is the employee's take-home pay.

The payroll envelope is kept by the employee. The payroll receipt is signed by the employee and kept by the employer.

As shown in Figure 2-2A, all three employees worked only on the Baker job during the pay period ending January 8. A.L. King and J.E. King each worked 17 hours. D.L. White worked 16 hours.

Figures 2-2B through 2-2J show additional examples of payroll receipts. Like the weekly time sheets, these receipts span a period of one year and cover three different jobs: the Baker, Brown and Green jobs.

Date	1-8-XX		Date	1-8-XX		Date	1-8-XX
Employee	ARNOLD L. KING		Employee	JERRY E. KING		Employee	DANIEL L. WHITE

Receipt 1 (Arnold L. King):
- Number of withholding exemptions: 1
- 17 Hours on BAKER job
- 17 Total hours @ 25.00 = $425.00
- Deductions:
  - Social Security $28.48
  - Federal withholding tax $62.40
  - State withholding tax $18.97
  - Other charges $____
  - Total Deductions $109.85
  - Net Amount $315.15

Receipt 2 (Jerry E. King):
- Number of withholding exemptions: 2
- 17 Hours on BAKER job
- 17 Total hours @ 25.00 = $425.00
- Deductions:
  - Social Security $28.48
  - Federal withholding tax $57.80
  - State withholding tax $18.31
  - Other charges $____
  - Total Deductions $104.59
  - Net Amount $320.41

Receipt 3 (Daniel L. White):
- Number of withholding exemptions: 1
- 16 Hours on BAKER job
- 16 Total hours @ 30.00 = $480.00
- Deductions:
  - Social Security $32.16
  - Federal withholding tax $77.20
  - State withholding tax $21.76
  - Other charges $____
  - Total Deductions $131.12
  - Net Amount $348.88

Payroll receipts
Figure 2-2A

Date __3 – 12 – XX__

Employee __DANIEL L. WHITE__

Number of withholding exemptions __1__

38½ Hours on __BROWN__ job

1½ Hours on __GREEN__ job

___ Hours on _____ job

40 Total hours @ 30.00 = $ 1200.00

Deductions:

Social Security $ 80.40

Federal withholding tax $ 327.70

State withholding tax $ 63.25

Other charges $ _____

    Total Deductions $ 471.35

    Net Amount $ 728.65

Employee Signature _____

Date __3 – 12 – XX__

Employee __ARNOLD L. KING__

Number of withholding exemptions __1__

32 Hours on __BROWN__ job

___ Hours on _____ job

___ Hours on _____ job

32 Total hours @ 25.00 = $ 800.00

Deductions:

Social Security $ 53.60

Federal withholding tax $ 181.50

State withholding tax $ 40.25

Other charges $ _____

    Total Deductions $ 275.35

    Net Amount $ 524.65

Employee Signature _____

Date __3 – 12 – XX__

Employee __JERRY E. KING__

Number of withholding exemptions __2__

24 Hours on __BROWN__ job

___ Hours on _____ job

___ Hours on _____ job

24 Total hours @ 25.00 = $ 600.00

Deductions:

Social Security $ 40.20

Federal withholding tax $ 104.90

State withholding tax $ 28.09

Other charges $ _____

    Total Deductions $ 173.19

    Net Amount $ 426.81

Employee Signature _____

Payroll receipts
Figure 2-2B

Date __4 – 16 – XX__

Employee __DANIEL L. WHITE__

Number of withholding exemptions __1__

40 Hours on __GREEN__ job

___ Hours on _____ job

___ Hours on _____ job

40 Total hours @ 30.00 = $ 1200.00

Deductions:

Social Security $ 80.40

Federal withholding tax $ 327.70

State withholding tax $ 63.25

Other charges $ _____

    Total Deductions $ 471.35

    Net Amount $ 728.65

Employee Signature _____

Date __4 – 16 – XX__

Employee __JERRY E. KING__

Number of withholding exemptions __2__

37 Hours on __BROWN__ job

3 Hours on __GREEN__ job

___ Hours on _____ job

40 Total hours @ 25.00 = $ 1000.00

Deductions:

Social Security $ 67.00

Federal withholding tax $ 246.50

State withholding tax $ 51.09

Other charges $ _____

    Total Deductions $ 364.59

    Net Amount $ 635.41

Employee Signature _____

Date __4 – 16 – XX__

Employee __DENNIS L. WEST__

Number of withholding exemptions __NONE__

32 Hours on __GREEN__ job

___ Hours on _____ job

___ Hours on _____ job

32 Total hours @ 25.00 = $ 800.00

Deductions:

Social Security $ 53.60

Federal withholding tax $ 221.00

State withholding tax $ 40.91

Other charges $ _____

    Total Deductions $ 315.51

    Net Amount $ 484.49

Employee Signature _____

Payroll receipts
Figure 2-2C

Date __6-25-XX__

Employee __DANIEL L. WHITE__

Number of withholding exemptions __1__

8 Hours on __BROWN__ job

19½ Hours on __GREEN__ job

___ Hours on _____ job

27½ Total hours @ 30.00 = $ __825.00__

Deductions:

Social Security $ __55.28__

Federal withholding tax $ __188.90__

State withholding tax $ __41.68__

Other charges $ _____

Total Deductions $ __285.86__

Net Amount $ __539.14__

Employee Signature _____

Date __6-25-XX__

Employee __CLARENCE A. LESTER__

Number of withholding exemptions __3__

19½ Hours on __GREEN__ job

___ Hours on _____ job

___ Hours on _____ job

19½ Total hours @ 30.00 = $ __585.00__

Deductions:

Social Security $ __39.20__

Federal withholding tax $ __93.80__

State withholding tax $ __26.55__

Other charges $ _____

Total Deductions $ __159.55__

Net Amount $ __425.45__

Employee Signature _____

Date __6-25-XX__

Employee __ARNOLD L. KING__

Number of withholding exemptions __1__

6½ Hours on __BROWN__ job

19½ Hours on __GREEN__ job

___ Hours on _____ job

26 Total hours @ 25.00 = $ __650.00__

Deductions:

Social Security $ __43.55__

Federal withholding tax $ __127.10__

State withholding tax $ __31.62__

Other charges $ _____

Total Deductions $ __202.27__

Net Amount $ __447.73__

Employee Signature _____

Payroll receipts
Figure 2-2D

Date __6-25-XX__

Employee __JERRY E. KING__

Number of withholding exemptions __2__

8 Hours on __BROWN__ job

19½ Hours on __GREEN__ job

___ Hours on _____ job

27½ Total hours @ 25.00 = $ __687.50__

Deductions:

Social Security $ __46.06__

Federal withholding tax $ __130.50__

State withholding tax $ __33.12__

Other charges $ _____

Total Deductions $ __209.68__

Net Amount $ __477.82__

Employee Signature _____

Date __6-25-XX__

Employee __DENNIS L. WEST__

Number of withholding exemptions __NONE__

19½ Hours on __GREEN__ job

___ Hours on _____ job

___ Hours on _____ job

19½ Total hours @ 25.00 = $ __487.50__

Deductions:

Social Security $ __32.66__

Federal withholding tax $ __105.90__

State withholding tax $ __23.08__

Other charges $ _____

Total Deductions $ __161.64__

Net Amount $ __325.86__

Employee Signature _____

Payroll receipts
Figure 2-2E

Date _8-6-XX_
Employee _ROBERT R. LEWIS_
Number of withholding exemptions _2_
_40_ Hours on _GREEN_ job
___ Hours on _____ job
___ Hours on _____ job
_40_ Total hours @ _30.00_ = $ _1200.00_

Deductions:
Social Security $ _80.40_
Federal withholding tax $ _320.50_
State withholding tax $ _62.59_
Other charges $ _____
Total Deductions $ _463.49_
Net Amount $ _736.51_

Employee Signature _____

Date _8-6-XX_
Employee _CLARENCE A. LESTER_
Number of withholding exemptions _3_
_40_ Hours on _GREEN_ job
___ Hours on _____ job
___ Hours on _____ job
_40_ Total hours @ _30.00_ = $ _1200.00_

Deductions:
Social Security $ _80.40_
Federal withholding tax $ _313.40_
State withholding tax $ _61.92_
Other charges $ _____
Total Deductions $ _455.72_
Net Amount $ _744.28_

Employee Signature _____

Date _8-6-XX_
Employee _ARNOLD L. KING_
Number of withholding exemptions _1_
_40_ Hours on _GREEN_ job
___ Hours on _____ job
___ Hours on _____ job
_40_ Total hours @ _25.00_ = $ _1000.00_

Deductions:
Social Security $ _67.00_
Federal withholding tax $ _253.70_
State withholding tax $ _51.75_
Other charges $ _____
Total Deductions $ _372.45_
Net Amount $ _627.55_

Employee Signature _____

Payroll receipts
Figure 2-2F

Date _8-6-XX_
Employee _JERRY E. KING_
Number of withholding exemptions _2_
_40_ Hours on _GREEN_ job
___ Hours on _____ job
___ Hours on _____ job
_40_ Total hours @ _25.00_ = $ _1000.00_

Deductions:
Social Security $ _67.00_
Federal withholding tax $ _246.50_
State withholding tax $ _51.09_
Other charges $ _____
Total Deductions $ _364.59_
Net Amount $ _635.41_

Employee Signature _____

Date _8-6-XX_
Employee _DENNIS L. WEST_
Number of withholding exemptions _NONE_
_36_ Hours on _GREEN_ job
___ Hours on _____ job
___ Hours on _____ job
_36_ Total hours @ _25.00_ = $ _900.00_

Deductions:
Social Security $ _60.30_
Federal withholding tax $ _258.00_
State withholding tax $ _46.66_
Other charges $ _____
Total Deductions $ _364.96_
Net Amount $ _535.04_

Employee Signature _____

Date _8-6-XX_
Employee _LAWSON H. KIDD_
Number of withholding exemptions _1_
_40_ Hours on _GREEN_ job
___ Hours on _____ job
___ Hours on _____ job
_40_ Total hours @ _25.00_ = $ _1000.00_

Deductions:
Social Security $ _67.00_
Federal withholding tax $ _287.90_
State withholding tax $ _51.75_
Other charges $ _____
Total Deductions $ _406.65_
Net Amount $ _593.35_

Employee Signature _____

Payroll receipts
Figure 2-2G

Date _____9 - 24 - x x_____

Employee _____CLARENCE A LESTER_____

Number of withholding exemptions _____3_____

32 Hours on _____GREEN_____ job

_____ Hours on _____ job

_____ Hours on _____ job

32 Total hours @ 30.00 = $ 960.00

Deductions:

Social Security $ 64.32

Federal withholding tax $ 224.60

State withholding tax $ 48.12

Other charges $ _____

          Total Deductions $ 337.04

          Net Amount $ 622.96

Employee Signature _____

Date _____9 - 24 - x x_____

Employee _____ARNOLD L. KING_____

Number of withholding exemptions _____1_____

16 Hours on _____GREEN_____ job

_____ Hours on _____ job

_____ Hours on _____ job

16 Total hours @ 25.00 = $ 400.00

Deductions:

Social Security $ 26.80

Federal withholding tax $ 57.60

State withholding tax $ 17.82

Other charges $ _____

          Total Deductions $ 102.22

          Net Amount $ 297.78

Employee Signature _____

Date _____9 - 24 - x x_____

Employee _____JERRY E. KING_____

Number of withholding exemptions _____2_____

16 Hours on _____GREEN_____ job

_____ Hours on _____ job

_____ Hours on _____ job

16 Total hours @ 25.00 = $ 400.00

Deductions:

Social Security $ 26.80

Federal withholding tax $ 53.00

State withholding tax $ 17.16

Other charges $ _____

          Total Deductions $ 96.96

          Net Amount $ 303.04

Employee Signature _____

**Payroll receipts
Figure 2-2H**

Date _____12 - 24 - x x_____

Employee _____DANIEL L. WHITE_____

Number of withholding exemptions _____1_____

40 Hours on _____GREEN_____ job

_____ Hours on _____ job

_____ Hours on _____ job

40 Total hours @ 30.00 = $ 1200.00

Deductions:

Social Security $ 80.40

Federal withholding tax $ 327.70

State withholding tax $ 63.25

Other charges $ _____

          Total Deductions $ 471.35

          Net Amount $ 728.65

Employee Signature _____

Date _____12 - 24 - x x_____

Employee _____ARNOLD L. KING_____

Number of withholding exemptions _____1_____

40 Hours on _____GREEN_____ job

_____ Hours on _____ job

_____ Hours on _____ job

40 Total hours @ 25.00 = $ 1000.00

Deductions:

Social Security $ 67.00

Federal withholding tax $ 253.70

State withholding tax $ 51.75

Other charges $ _____

          Total Deductions $ 372.45

          Net Amount $ 627.55

Employee Signature _____

Date _____12 - 24 - x x_____

Employee _____JERRY E. KING_____

Number of withholding exemptions _____2_____

40 Hours on _____GREEN_____ job

_____ Hours on _____ job

_____ Hours on _____ job

40 Total hours @ 25.00 = $ 1000.00

Deductions:

Social Security $ 67.00

Federal withholding tax $ 246.50

State withholding tax $ 51.09

Other charges $ _____

          Total Deductions $ 364.59

          Net Amount $ 635.41

Employee Signature _____

**Payroll receipts
Figure 2-2I**

Date ___12-31-XX___

Employee ___ARNOLD L. KING___

Number of withholding
exemptions ___1___

_16_ Hours on ___GREEN___ job

___ Hours on _____ job

___ Hours on _____ job

_16_ Total hours @ _25.00_ = $ _400.00_

Deductions:

Social Security $ _26.80_

Federal withholding tax $ _57.60_

State withholding tax $ _17.82_

Other charges $ _____

Total Deductions $ _102.22_

Net Amount $ _297.78_

Employee
Signature _____

Date ___12-31-XX___

Employee ___JERRY E. KING___

Number of withholding
exemptions ___2___

_32_ Hours on ___GREEN___ job

___ Hours on _____ job

___ Hours on _____ job

_32_ Total hours @ _25.00_ = $ _800.00_

Deductions:

Social Security $ _53.60_

Federal withholding tax $ _174.40_

State withholding tax $ _39.59_

Other charges $ _____

Total Deductions $ _267.59_

Net Amount $ _532.41_

Employee
Signature _____

**Payroll receipts
Figure 2-2J**

Individual Employee Earnings Record

The information recorded on the payroll envelope and receipt is transferred to the individual employee earnings record. This will give you a permanent record of the employee's earnings and deductions. You'll use this record to prepare your quarterly and year-end reports for the federal and state governments.

Look at Figure 2-3A. A.L. King didn't work the first week in January. For the second week in January, A.L. King's gross pay was $425.00. This figure is entered in the total earnings column for week number 2.

His deductions came to: $28.48 for social security, $62.40 for federal withholding tax, and $18.97 for state withholding tax. Each of these amounts is entered in the correct column.

If there were other deductions, they would be entered in the extra columns shown in this section. A.L. King's net pay (take-home pay) was $315.15.

Now look at Figure 2-3B. J.E. King also had gross earnings of $425.00 for the week ending January 8. But his total deductions were not the same as those for A.L. King. This is because J.E. King claimed two exemptions instead of one.

His social security deduction was the same ($28.48), because social security deductions are not affected by the number of exemptions claimed. But the number of exemptions does determine the federal and state withholding taxes. J.E. King's federal withholding tax was $57.80. His state withholding tax was $18.31. His net pay for the pay period was $320.41.

As shown in Figure 2-3C, D.L. White had one exemption. His gross earnings for week number 2 were $480.00. His social security deduction was $32.16. His federal withholding tax was $77.20. His state withholding tax was $21.76. White's net pay was $348.88.

Make sure that every week is accounted for. In Figure 2-3C, D.L. White didn't work for several weeks. His absence was indicated by a series of dashes. It's important to put in these dashes. If the dashes were omitted, it might look like an entry had been forgotten.

Notice that the week ending March 26 was the last payroll period for the first quarter. Add up each column and enter the totals on the "Total 1st Qtr" line. Be sure to recheck all totals before preparing any quarterly reports.

NAME OF EMPLOYEE	ARNOLD L. KING			SOCIAL SECURITY NUMBER	000 - 00 - 0000	
ADDRESS				CITY OR TOWN		
DATE OF BIRTH		MARRIED ☒ OR SINGLE ☐	NUMBER OF EXEMPTIONS 1	PHONE NO.	CLOCK NO.	
POSITION CARPENTER HELPER	RATE $25.00/HR	DATE 5-2-XX	DATE STARTED 6-07-XX	DATE TERMINATED		
REMARKS				REASON		

FIRST QUARTER 19 XX

WEEK #	DATE	HOURS WORKED REG.	OVER-TIME	TOTAL EARNINGS	SOC. SEC.	FED. WITH. TAX	STATE WITH. TAX			NET PAY
1	1/1	—	—	—	—	—	—			—
2	1/8	17	—	425 00	28.48	62 40	18.97			315 15
3	1/15	—	—	—	—	—	—			—
4	1/22	—	—	—	—	—	—			—
5	1/29	—	—	—	—	—	-			—
6	2/5	—	—	—	—	—	—			—
7	2/12	—	—	—	—	—	—			—
8	2/19	—	—	—	—	—	—			—
9	2/26	—	—	—	—	—	—			—
10	3/5	—	—	—	—	—	—			—
11	3/12	32	—	800 00	53.60	181 50	40.25			524 65
12	3/19	40	—	1000 00	67.00	253 70	51.75			627 55
13	3/26	40	—	1000 00	67.00	253 70	51.75			627 55
Total 1st Qtr				3225 00	216.08	751 30	162.72			2094 90
TOTAL 3 MOS				3225 00	216.08	751 30	162.72			2094 90

SECOND QUARTER 19 XX

WEEK #	DATE	HOURS WORKED REG.	OVER-TIME	TOTAL EARNINGS	SOC. SEC.	FED. WITH. TAX	STATE WITH. TAX			NET PAY
14	4/2	40	—	1000 00	67.00	253 70	51.75			627 55
15	4/9	40	—	1000 00	67.00	253 70	51.75			627 55
16	4/16	—	—	—	—	—	—			—
17	4/23	32	—	800 00	53.60	181 50	40.25			524 65
18	4/30	39½	—	987 50	66.16	249 08	51.03			621 23
19	5/7	31	—	775 00	51.93	170 40	38.81			513 86
20	5/14	32	—	800 00	53.60	181 50	40.25			524 65
21	5/21	40	—	1000 00	67.00	253 70	51.75			627 55
22	5/28	39	—	975 00	65.33	244 45	50.31			614 91
23	6/4	30	—	750 00	50.25	163 00	37.57			499 38
24	6/11	18	—	450 00	30.15	69 60	20.69			329 56
25	6/18	32	—	800 00	53.60	181 50	40.25			524 65
26	6/25	26	—	650 00	43.55	127 10	31.62			447 73
Total 2nd Qtr				9,987 50	669.17	2329 23	505.83			6,483 27
TOTAL 6 MOS				13,212 50	885.25	3080 53	668.55			8,578 17

THIRD QUARTER 19 XX

WEEK #	DATE	HOURS WORKED REG.	OVER-TIME	TOTAL EARNINGS	SOC. SEC.	FED. WITH. TAX	STATE WITH. TAX			NET PAY
27	7/2	40	—	1000 00	67.00	253 70	51.75			627 55
28	7/6	32	—	800 00	53.60	181 50	40.25			524 65
29	7/16	40	-	1000 00	67.00	253 70	51.75			627 55
30	7/23	—	—	—	—	—	—			—
31	7/30	—	—	—	—	—	—			—
32	8/6	40	—	1000 00	67.00	253 70	51.75			627 55
33	8/13	40	—	1000 00	67.00	253 70	51.75			627 55
34	8/20	39	—	975 00	65.33	244 45	50.31			614 91
35	8/27	40	—	1000 00	67.00	253 70	51.75			627 55
36	9/3	40	—	1000 00	67.00	253 70	51.75			627 55
37	9/10	32	—	800 00	53.60	181 50	40.25			524 65
38	9/17	40	—	1000 00	67.00	253 70	51.75			627 55
39	9/24	16	—	400 00	26.80	57 60	17.82			297 78
Total 3rd Qtr				9,975 00	668.33	2,440 95	510.88			6,354 84
TOTAL 9 MOS				23,187 50	1,563.58	5521 48	1179.43			14,933 01

FOURTH QUARTER 19 XX

WEEK #	DATE	HOURS WORKED REG.	OVER-TIME	TOTAL EARNINGS	SOC. SEC.	FED. WITH. TAX	STATE WITH. TAX			NET PAY
40	10/1	40	—	1000 00	67.00	253 70	51.75			627 55
41	10/8	40	-	1000 00	67.00	253 70	51.75			627 55
42	10/15	40	—	1000 00	67.00	253 70	51.75			627 55
43	10/22	25½	—	637 50	42.71	120 70	30.90			443 19
44	10/29	32	—	800 00	53.60	181 50	40.25			524 65
45	11/5	40	—	1000 00	67.00	253 70	51.75			627 55
46	11/12	40	—	1000 00	67.00	253 70	51.75			627 55
47	11/19	32	—	800 00	53.60	181 50	40.25			524 65
48	11/26	24	—	600 00	40.20	111 10	28.75			419 95
49	12/3	40	—	1000 00	67.00	253 70	51.75			627 55
50	12/10	32	—	800 00	53.60	181 50	40.25			524 65
51	12/17	40	—	1000 00	67.00	253 70	51.75			627 55
52	12/24	40	—	1000 00	67.00	253 70	51.75			627 55
	12/31	16	—	400 00	26.80	57 60	17.82			297 78
Total 4th Qtr				12,037 50	806.51	2863 50	612.72			7,755 27
TOTAL YEAR				35,225 00	2,360.09	8384 98	1791.65			22,688 28

Individual employee earnings record
Figure 2-3A

NAME OF EMPLOYEE	JERRY E. KING	SOCIAL SECURITY NUMBER	000-00-0000
ADDRESS		CITY OR TOWN	
DATE OF BIRTH	MARRIED ☒ OR SINGLE ☐ NUMBER OF EXEMPTIONS 2	PHONE NO.	CLOCK NO.
POSITION CARPENTER HELPER RATE $25.00 DATE 5-2-XX		DATE STARTED 6-7-XX DATE TERMINATED	
REMARKS		REASON	

FIRST QUARTER 19XX

WEEK #	DATE	REG.	OVER TIME	TOTAL EARNINGS	SOC. SEC.	FED. WITH. TAX	STATE WITH. TAX			NET PAY
1	1/1	—	—	—	—	—	—			—
2	1/8	17	—	425 00	28.48	57 80	18.31			320 41
3	1/15	—	—	—	—	—	—			—
4	1/22	—	—	—	—	—	—			—
5	1/29	—	—	—	—	—	—			—
6	2/5	—	—	—	—	—	—			—
7	2/12	—	—	—	—	—	—			—
8	2/19	—	—	—	—	—	—			—
9	2/26	—	—	—	—	—	—			—
10	3/5	—	—	—	—	—	—			—
11	3/12	24	—	600 00	40.20	104 90	28.09			426 81
12	3/19	40	—	1000 00	67.00	246 50	51.09			635 41
13	3/26	40	—	1000 00	67.00	246 50	51.09			635 41
Total 1st Qtr				3025 00	202.68	655 70	148.58			2018 04
TOTAL 3 MOS				3025 00	202.68	655.70	148.58			2018 04

SECOND QUARTER 19XX

WEEK #	DATE	REG.	OVER TIME	TOTAL EARNINGS	SOC. SEC.	FED. WITH. TAX	STATE WITH. TAX			NET PAY
14	4/2	40	—	1000 00	67.00	246 50	51.09			635 41
15	4/9	40	—	1000 00	67.00	246 50	51.09			635 41
16	4/16	40	—	1000 00	67.00	246 50	51.09			635 41
17	4/23	30	—	750 00	50.25	155 90	36.71			507 14
18	4/30	40	—	1000 00	67.00	246 50	51.09			635 41
19	5/7	31	—	775 00	51.93	163 30	38.15			521 62
20	5/14	32	—	800 00	53.60	174 40	39.59			532 41
21	5/21	40	—	1000 00	67.00	246 50	51.09			635 41
22	5/28	39	—	975 00	65.33	237 25	49.65			622 77
23	6/4	36½	—	912 50	61.14	214 13	46.05			591 18
24	6/11	18	—	450 00	30.15	65 00	20.03			334 82
25	6/18	32	—	800 00	53.60	174 40	39.59			532 41
26	6/25	27½	—	687 50	46.06	130 50	33.12			477 82
Total 2nd Qtr				11,150 00	747.06	2547.38	558.34			7,297 22
TOTAL 6 MOS				14,175 00	949.74	3,203.08	706.92			9,315 26

THIRD QUARTER 19XX

WEEK #	DATE	REG.	OVER TIME	TOTAL EARNINGS	SOC. SEC.	FED. WITH. TAX	STATE WITH. TAX			NET PAY
27	7/2	40	—	1000 00	67.00	246 50	51.09			635 41
28	7/9	32	—	800 00	53.60	174 40	39.59			532 41
29	7/16	40	—	1000 00	67.00	246 50	51.09			635 41
30	7/23	—	—	—	—	—	—			—
31	7/30	—	—	—	—	—	—			—
32	8/6	40	—	1000 00	67.00	246 50	51.09			635 41
33	8/13	40	—	1000 00	67.00	246 50	51.08			635 41
34	8/20	39	—	975 00	65.33	237 25	49.65			622 77
35	8/27	40	—	1000 00	67.00	246 50	51.09			635 41
36	9/3	40	—	1000 00	67.00	246 50	51.09			635 41
37	9/10	32	—	800 00	53.60	174 40	39.59			532 41
38	9/17	32	—	800 00	53.60	174 40	39.59			532 41
39	9/24	16	—	400 00	26.80	53 00	17.16			303 04
Total 3rd Qtr				9,775 00	654.93	2,792 45	492.12			6,335 50
TOTAL 9 MOS				23,950 00	1604.87	5,495 53	1,179.04			15,650 76

FOURTH QUARTER 19

WEEK #	DATE	REG.	OVER TIME	TOTAL EARNINGS	SOC. SEC.	FED. WITH. TAX	STATE WITH. TAX			NET PAY
40	10/1	40	—	1000 00	67.00	246 50	51.09			635 41
41	10/8	40	—	1000 00	67.00	246 50	51.09			635 41
42	10/15	40	—	1000 00	67.00	246 50	51.09			635 41
43	10/22	33½	—	837 50	56.11	185 50	41.74			554 15
44	10/29	16	—	400 00	26.80	53 00	17.16			303 04
45	11/5	40	—	1000 00	67.00	246 50	51.09			635 41
46	11/12	40	—	1000 00	67.00	246 50	51.09			635 41
47	11/19	32	—	800 00	53.60	174 40	39.59			532 41
48	11/26	24	—	600 00	40.20	104 90	28.09			426 81
49	12/3	40	—	1000 00	67.00	246 50	51.09			635 41
50	12/10	32	—	800 00	53.60	174 40	39.59			532 41
51	12/17	40	—	1000 00	67.00	246 50	51.09			635 41
52	12/24	40	—	1000 00	67.00	246 50	51.09			635 41
	12/31	32	—	800 00	53.60	174 40	39.59			532 41
Total 4th Qtr				12,237 50	819.91	2,838.60	614.48			7,964 51
TOTAL YEAR				36,187 50	2,424.58	8,334.13	1,813.52			23,615 27

Individual employee earnings record
Figure 2-3B

NAME OF EMPLOYEE	DANIEL L. WHITE					SOCIAL SECURITY NUMBER	000-00-0000		
ADDRESS						CITY OR TOWN			

DATE OF BIRTH		MARRIED ☒ OR SINGLE ☐	NUMBER OF EXEMPTIONS 1		PHONE NO.		CLOCK NO.	
POSITION CARPENTER	RATE $30.00	DATE 5-2-XX			DATE STARTED 12-14-XX		DATE TERMINATED 7-11-XX	
REMARKS					REASON			

FIRST QUARTER 19XX

WEEK #	DATE	REG.	OVER TIME	TOTAL EARNINGS	SOC. SEC.	FED. WITH. TAX	STATE WITH. TAX			NET PAY
1	1/1	—	—							
2	1/8	16	—	480 00	32.16	77 20	21.76			348 88
3	1/15	4	—	120 00	8.04	6 90	2.44			102 57
4	1/22	9½	—	285 00	19.10	33 40	10.92			221 58
5	1/29	8	—	240 00	16.08	25 90	8.68			189 34
6	2/5	—	—							
7	2/12	24	—	720 00	48.24	151 90	35.65			484 21
8	2/19	32	—	960 00	64.32	238 90	49.25			607 33
9	2/26	—	—							
10	3/5	—	—							
11	3/12	40	—	1200 00	80.40	327 70	63.25			728 65
12	3/19	40	—	1200 00	80.40	327 70	63.25			728 65
13	3/26	40	—	1200 00	80.40	327 70	63.25			728 65
Total 1st Qtr				6405 00	429.14	1517.30	318.70			4139 86
TOTAL 3 MOS				6405 00	429.14	1517.30	318.70			4139 86

SECOND QUARTER 19XX

WEEK #	DATE	REG.	OVER TIME	TOTAL EARNINGS	SOC. SEC.	FED. WITH. TAX	STATE WITH. TAX			NET PAY
14	4/2	40	—	1200 00	80.40	327 70	63.25			728 65
15	4/9	40	—	1200 00	80.40	327 70	63.25			728 65
16	4/16	40	—	1200 00	80.40	327 70	63.25			728 65
17	4/23	37	—	1110 00	74.37	294 40	58.07			683 16
18	4/30	40	—	1200 00	80.40	327 70	63.25			728 65
19	5/7	31	—	930 00	62.31	227 80	47.72			592 17
20	5/14	40	—	1200 00	80.40	327 70	63.25			728 65
21	5/21	40	—	1200 00	80.40	327 70	63.25			728 65
22	5/28	15	—	450 00	30.15	69 60	20.69			329 56
23	6/4	30	—	900 00	60.30	216 70	46.00			577 00
24	6/11	15	—	450 00	30.15	69 60	20.69			329 56
25	6/18	25½	—	765 00	51.26	166 70	38.23			508 81
26	6/25	27½	—	825 00	55.28	188 90	41.68			539 14
Total 2nd Qtr				12,630 00	846.22	3199 90	652.58			7931 30
TOTAL 6 MOS				19,035 00	1275.36	4717.20	971.28			12071 16

THIRD QUARTER 19XX

WEEK #	DATE	REG.	OVER TIME	TOTAL EARNINGS	SOC. SEC.	FED. WITH. TAX	STATE WITH. TAX			NET PAY
27	7/2	40	—	1200 00	80.40	327 70	63.25			728 65
28	7/9	—	—							
29	7/16	—	—							
30	7/23	—	—							
31	7/30	—	—							
32	8/6	—	—							
33	8/13	—	—							
34	8/20	—	—							
35	8/27	—	—							
36	9/3	—	—							
37	9/10	—	—							
38	9/17	—	—			—				
39	9/24	—	—			—				
				1200 00	80.40	327 70	63.25			728 65
Total 3rd Qtr				20,235 00	1355.76	5044.90	1034.63			12799 81
TOTAL 9 MOS										

FOURTH QUARTER 19XX

WEEK #	DATE	REG.	OVER TIME	TOTAL EARNINGS	SOC. SEC.	FED. WITH. TAX	STATE WITH. TAX			NET PAY
40										
41										
42										
43										
44										
45										
46										
47										
48										
49										
50										
51	12/17	20½	—	615 00	41.21	114 30	29.61			429 88
52	12/24	40	—	1200 00	80.40	327 70	63.25			728 65
	12/31	—	—			—				
Total 4th Qtr				1,815 00	121.61	442 00	92.86			1,158 53
TOTAL YEAR				22,050 00	1477.57	5486 90	1127.39			13958 34

Individual employee earnings record
Figure 2-3C

NAME OF EMPLOYEE	DENNIS L. WEST		SOCIAL SECURITY NUMBER	000 - 00 - 0000	
ADDRESS			CITY OR TOWN		
DATE OF BIRTH	MARRIED ☐ OR SINGLE ☒	NUMBER OF EXEMPTIONS NONE	PHONE NO.		CLOCK NO.
POSITION CARPENTER HELPER	RATE $25.00	DATE 4-12-XX	DATE STARTED 4-12-XX	DATE TERMINATED 9-17-XX	
REMARKS			REASON		

FIRST QUARTER 19 / SECOND QUARTER 19

WEEK #	DATE	HOURS WORKED REG.	OVER-TIME	TOTAL EARNINGS	SOC. SEC.	FED. WITH. TAX	STATE WITH. TAX			NET PAY		WEEK #	DATE	HOURS WORKED REG.	OVER-TIME	TOTAL EARNINGS	SOC. SEC.	FED. WITH. TAX	STATE WITH. TAX			NET PAY	
1												14											
2												15											
3												16	4/16	32	—	800 00	53.60	221 00	40.91			484	49
4												17	4/23	37	—	925 00	61.98	267 25	48.09			547	68
5												18	4/30	32	—	800 00	53.60	221 00	40.91			484	49
6												19	5/7	31	—	775 00	51.93	211 75	39.47			471	85
7												20	5/14	40	—	1000 00	67.00	295 00	52.41			585	59
8												21	5/21	36	—	900 00	60.30	258 00	46.66			535	04
9												22	5/28	31	—	775 00	51.93	211 75	39.47			471	85
10												23	6/4	16	—	400 00	26.80	79 80	18.48			274	92
11												24	6/11	—	—	—		—					
12												25	6/18	—	—	—		—					
13												26	6/25	19½	—	487 50	32.66	105 90	23.08			325	86
Total 1st Qtr												Total 2nd Qtr				6862 50	459.80	1871.45	349.48			4,181	77
TOTAL 3 MOS												TOTAL 6 MOS				6862 50	459.80	1871.45	349.48			4,181	77

THIRD QUARTER 19 / FOURTH QUARTER 19

WEEK #	DATE	HOURS WORKED REG.	OVER-TIME	TOTAL EARNINGS	SOC. SEC.	FED. WITH. TAX	STATE WITH. TAX			NET PAY		WEEK #	DATE	HOURS WORKED REG.	OVER-TIME	TOTAL EARNINGS	SOC. SEC.	FED. WITH. TAX	STATE WITH. TAX			NET PAY	
27	7/2	40	—	1000 00	67.00	295 00	52.41			585	59	40											
28	7/9	32	—	800 00	53.60	221 00	40.91			484	49	41											
29	7/16	32	—	800 00	53.60	221 00	40.91			484	49	42											
30	7/23	40	—	1000 00	67.00	295 00	52.41			585	59	43											
31	7/30	24	—	600 00	40.20	148 90	29.91			381	49	44											
32	8/6	36	—	900 00	60.30	258 00	46.66			535	04	45											
33	8/13	40	—	1000 00	67.00	295 00	52.41			585	59	46											
34	8/20	15	—	375 00	25.13	70 80	16.76			262	31	47											
35	8/27	32	—	800 00	53.60	221 00	40.91			484	49	48											
36	9/3	16	—	400 00	26.80	79 80	18.48			274	92	49											
37	9/10	32	—	800 00	53.60	221 00	40.91			484	49	50											
38	9/17	24	—	600 00	40.20	148 90	29.41			381	49	51											
39												52											
Total 3rd Qtr				9075.00	608.03	2475.40	461.59			5529	98	Total 4th Qtr				—	—	—	—			—	
TOTAL 9 MOS				15937.50	1067.83	4346.85	811.07			9,711	75	TOTAL YEAR				15,937 50	1,067.83	4346 85	811.07			9,711	75

Individual employee earnings record
Figure 2-3D

NAME OF EMPLOYEE	CLARENCE A. LESTER				SOCIAL SECURITY NUMBER		000-00-0000	
ADDRESS					CITY OR TOWN			
DATE OF BIRTH		MARRIED ☒ OR SINGLE ☐	NUMBER OF EXEMPTIONS 3		PHONE NO.		CLOCK NO.	
POSITION CARPENTER	RATE $ 30.00	DATE 6-22-XX		DATE STARTED 6-22-XX		DATE TERMINATED 9-24-XX		
REMARKS					REASON			

FIRST QUARTER 19 XX

WEEK #	DATE	HOURS WORKED REG.	HOURS WORKED OVER-TIME	TOTAL EARNINGS	DEDUCTIONS SOC. SEC.	DEDUCTIONS FED. WITH. TAX	DEDUCTIONS STATE WITH. TAX			NET PAY	
1											
2											
3											
4											
5											
6											
7											
8											
9											
10											
11											
12											
13											
Total 1st Qtr											
TOTAL 3 MOS											

SECOND QUARTER 19 XX

WEEK #	DATE	HOURS WORKED REG.	HOURS WORKED OVER-TIME	TOTAL EARNINGS	DEDUCTIONS SOC. SEC.	DEDUCTIONS FED. WITH. TAX	DEDUCTIONS STATE WITH. TAX			NET PAY	
14											
15											
16											
17											
18											
19											
20											
21											
22											
23											
24											
25											
26	6/25	19½	—	585 00	39.20	93.80	26.55			425	45
Total 2nd Qtr				585 00	39.20	93.80	26.55			425	45
TOTAL 6 MOS				585 00	39.20	93.80	26.55			425	45

THIRD QUARTER 19 XX

WEEK #	DATE	HOURS WORKED REG.	HOURS WORKED OVER-TIME	TOTAL EARNINGS	DEDUCTIONS SOC. SEC.	DEDUCTIONS FED. WITH. TAX	DEDUCTIONS STATE WITH. TAX			NET PAY	
27	7/2	40	—	1200 00	80.40	313 40	61.92			744	28
28	7/9	24	—	720 00	48.24	137 70	34.32			499	74
29	7/16	—	—	—	—	—	—			—	
30	7/23	40	—	1200 00	80.40	313 40	61.92			744	28
31	7/30	40	—	1200 00	80.40	313 40	61.92			744	28
32	8/6	40	—	1200 00	80.40	313 40	61.92			744	28
33	8/13	40	—	1200 00	80.40	313 40	61.92			744	28
34	8/20	39	—	1170 00	78.39	302 30	60.19			729	12
35	8/27	40	—	1200 00	80.40	313 40	61.92			744	28
36	9/3	40	—	1200 00	80.40	313 40	61.92			744	28
37	9/10	32	—	960 00	64.32	224 60	48.12			622	96
38	9/17	40	—	1200 00	80.40	313 40	61.92			744	28
39	9/24	32	—	960 00	64.32	224 60	48.12			622	96
Total 3rd Qtr				13,410 00	898.47	3396.40	686.11			8,429	02
TOTAL 9 MOS				13,995 00	937.67	3490.20	712.66			8854	47

FOURTH QUARTER 19 XX

WEEK #	DATE	HOURS WORKED REG.	HOURS WORKED OVER-TIME	TOTAL EARNINGS	DEDUCTIONS SOC. SEC.	DEDUCTIONS FED. WITH. TAX	DEDUCTIONS STATE WITH. TAX			NET PAY	
40											
41											
42											
43											
44											
45											
46											
47											
48											
49											
50											
51											
52											
Total 4th Qtr		—	—	—	—	—	—			—	
TOTAL YEAR				13,995 00	937.67	3490.20	712.66			8854	47

Individual employee earnings record
Figure 2-3E

NAME OF EMPLOYEE	ROBERT R. LEWIS			SOCIAL SECURITY NUMBER	000 - 00 - 0000	
ADDRESS				CITY OR TOWN		
DATE OF BIRTH		MARRIED ☒ OR SINGLE ☐	NUMBER OF EXEMPTIONS 2	PHONE NO.		CLOCK NO.
POSITION CARPENTER	RATE $30.00	DATE 7-12-XX		DATE STARTED 7-12-XX	DATE TERMINATED 8-8-XX	
REMARKS				REASON		

FIRST QUARTER 19 XX

WEEK #	DATE	HOURS WORKED REG.	OVER-TIME	TOTAL EARNINGS	SOC. SEC.	FED. WITH. TAX	STATE WITH. TAX		NET PAY
1									
2									
3									
4									
5									
6									
7									
8									
9									
10									
11									
12									
13									
Total 1st Qtr									
TOTAL 3 MOS									

SECOND QUARTER 19 XX

WEEK #	DATE	HOURS WORKED REG.	OVER-TIME	TOTAL EARNINGS	SOC. SEC.	FED. WITH. TAX	STATE WITH. TAX		NET PAY
14									
15									
16									
17									
18									
19									
20									
21									
22									
23									
24									
25									
26									
Total 2nd Qtr									
TOTAL 6 MOS									

THIRD QUARTER 19 XX

WEEK #	DATE	HOURS WORKED REG.	OVER-TIME	TOTAL EARNINGS	SOC. SEC.	FED. WITH. TAX	STATE WITH. TAX		NET PAY
27									
28									
29	7/16	32	—	960 00	64.32	231 70	48.79		615 19
30	7/23	32	—	960 00	64.32	231 70	48.79		615 19
31	7/30	—	—	—	—	—	—		—
32	8/6	40	—	1200 00	80 40	320 50	62.59		736 51
33	8/13	8	—	240 00	16.08	22 80	8.10		193 02
34									
35									
36									
37									
38									
39									
Total 3rd Qtr				3,360 00	225.12	806 70	168.27		2,159 91
TOTAL 9 MOS				3,360 00	225.12	806 70	168.27		2,159 91

FOURTH QUARTER 19 XX

WEEK #	DATE	HOURS WORKED REG.	OVER-TIME	TOTAL EARNINGS	SOC. SEC.	FED. WITH. TAX	STATE WITH. TAX		NET PAY
40									
41									
42									
43									
44									
45									
46									
47									
48									
49									
50									
51									
52									
Total 4th Qtr		—	—	—	—	—	—		—
TOTAL YEAR				3360 00	225.12	806.70	168.27		2,159 91

Individual employee earnings record
Figure 2-3F

NAME OF EMPLOYEE	LAWSON H. KIDD		SOCIAL SECURITY NUMBER	000 - 00 - 0000	
ADDRESS			CITY OR TOWN		
DATE OF BIRTH		MARRIED ☐ OR SINGLE ☒	NUMBER OF EXEMPTIONS 1	PHONE NO.	CLOCK NO.
POSITION CARPENTER HELPER	RATE $25.00	DATE 7-18-XX	DATE STARTED 7-18-XX	DATE TERMINATED 11-18-XX	
REMARKS			REASON		

FIRST QUARTER 19 XX

WEEK #	DATE	HOURS WORKED REG.	OVER TIME	TOTAL EARNINGS	SOC. SEC.	FED. WITH. TAX	STATE WITH. TAX			NET PAY
1										
2										
3										
4										
5										
6										
7										
8										
9										
10										
11										
12										
13										
Total 1st Qtr										
TOTAL 3 MOS										

SECOND QUARTER 19 XX

WEEK #	DATE	HOURS WORKED REG.	OVER TIME	TOTAL EARNINGS	SOC. SEC.	FED. WITH. TAX	STATE WITH. TAX			NET PAY
14										
15										
16										
17										
18										
19										
20										
21										
22										
23										
24										
25										
26										
Total 2nd Qtr										
TOTAL 6 MOS										

THIRD QUARTER 19 XX

WEEK #	DATE	HOURS WORKED REG.	OVER TIME	TOTAL EARNINGS	SOC. SEC.	FED. WITH. TAX	STATE WITH. TAX			NET PAY
27										
28										
29										
30	7/23	40	—	1000 00	67.00	287 90	51.75			593 35
31	7/30	32	—	800 00	53.60	213 90	40.25			492 25
32	8/6	40	—	1000 00	67.00	287 90	51.75			593 35
33	8/13	—	—	—	—	—	—			—
34	8/20	—	—	—	—	—	—			—
35	8/27	—	—	—	—	—	—			—
36	9/3	—	—	—	—	—	—			—
37	9/10	—	—	—	—	—	—			—
38	9/17	—	—	—	—	—	—			—
39	9/24	—	—	—	—	—	—			—
Total 3rd Qtr				2,800 00	187.60	789.70	143.75			1678 95
TOTAL 9 MOS				2,800 00	187.60	789.70	143.75			1678 95

FOURTH QUARTER 19 XX

WEEK #	DATE	HOURS WORKED REG.	OVER TIME	TOTAL EARNINGS	SOC. SEC.	FED. WITH. TAX	STATE WITH. TAX			NET PAY
40										
41										
42										
43										
44										
45										
46	11/12	24	—	600 00	40.20	141 70	28.75			389 35
47	11/19	16	—	400 00	26.80	74 10	17.82			281 28
48										
49										
50										
51										
52										
Total 4th Qtr				1000 00	67.00	215 80	46.57			670 63
TOTAL YEAR				3,800 00	254.60	1005.50	190.32			2,349 58

Individual employee earnings record
Figure 2-3G

An effective cross-check is: the employee's total net pay for the quarter, plus his total deductions, must equal his total gross earnings. In Figure 2-3C, for example, D.L. White's gross earnings for the first quarter were $6,405.00. His net pay, plus his deductions, were:

Net pay	$4,139.86
State withholding tax	318.70
Federal withholding tax	1,517.30
Social security (FICA)	429.14
Total	$6,405.00

Does the net pay, plus deductions, equal the total gross earnings? Yes. Our arithmetic is correct. *Perform this check for each employee before preparing your quarterly reports.*

Figures 2-3D through 2-3G show additional examples of individual employee earnings records.

Again, these records span a period of one year and cover three different jobs: the Baker, Brown and Green jobs.

Consolidated Report of Labor Costs per Job

Figure 2-4A shows a consolidated report of labor costs per job. It's an optional report, but I recommend that you include it in your payroll record system. Here's why.

1) It shows you at a glance what your labor costs are for every job, even while the job is still in progress.

2) It's a cross-check against errors in your quarterly and year-end reports.

3) It's useful when paying the premiums on worker's compensation and liability insurance for your employees.

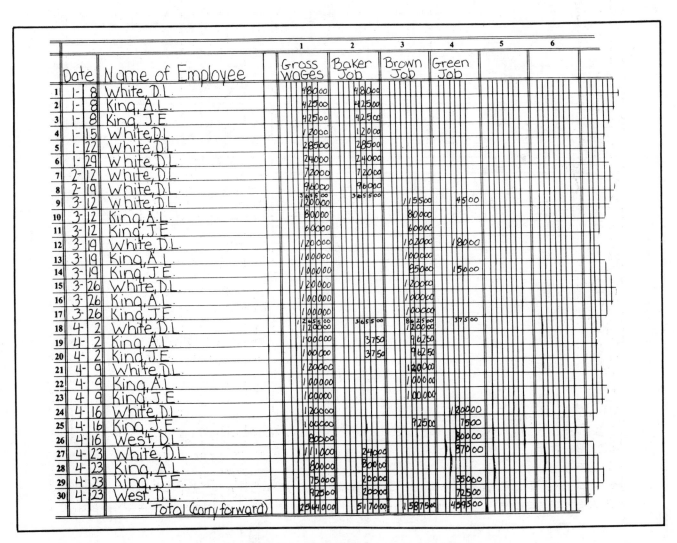

	Date	Name of Employee	1 Gross Wages	2 Baker Job	3 Brown Job	4 Green Job	5	6
1	1-8	White, D.L.	480.00	480.00				
2	1-8	King, A.L.	425.00	425.00				
3	1-8	King, J.E.	425.00	425.00				
4	1-15	White, D.L.	120.00	120.00				
5	1-22	White, D.L.	285.00	285.00				
6	1-29	White, D.L.	240.00	240.00				
7	2-12	White, D.L.	720.00	720.00				
8	2-19	White, D.L.	960.00	960.00				
9	3-12	White, D.L.	3655.00 / 1200.00	3655.00	1155.00	45.00		
10	3-12	King, A.L.	800.00		800.00			
11	3-12	King, J.E.	600.00		600.00			
12	3-19	White, D.L.	1200.00		1020.00	180.00		
13	3-19	King, A.L.	1000.00		1000.00			
14	3-19	King, J.E.	1000.00		850.00	150.00		
15	3-26	White, D.L.	1200.00		1200.00			
16	3-26	King, A.L.	1000.00		1000.00			
17	3-26	King, J.E.	1000.00		1000.00			
18	4-2	White, D.L.	12455.00 / 1200.00	3655.00	8625.00	375.00		
19	4-2	King, A.L.	1000.00	37.50	962.50			
20	4-2	King, J.E.	1000.00	37.50	962.50			
21	4-9	White, D.L.	1200.00		1200.00			
22	4-9	King, A.L.	1000.00		1000.00			
23	4-9	King, J.E.	1000.00		1000.00			
24	4-16	White, D.L.	1200.00			1200.00		
25	4-16	King, J.E.	1000.00		925.00	75.00		
26	4-16	West, D.L.	800.00			800.00		
27	4-23	White, D.L.	1110.00	240.00		870.00		
28	4-23	King, A.L.	800.00	800.00				
29	4-23	King, J.E.	750.00	200.00		550.00		
30	4-23	West, D.L.	925.00	200.00		725.00		
		Total (carry forward)	25440.00	5170.00	5875.00	4995.00		

Consolidated report of labor costs per job
Figure 2-4A

#	Date	Name of Employee	Gross Wages (1)	Baker Job (2)	Brown Job (3)	Green Job (4)	5	6	7	8
1		(Brought forward)	2564000	517000	1587500	459500				
2	4-30	White, D.L.	120000	43500	52500	24000				
3	4-30	King, A.L.	48750	30000	18750					
4	4-30	King, J.E.	100000	36250	43750	20000				
5	4-30	West, D.L.	80000	31250	28750	20000				
6	5-7	White, D.L.	2462750 / 93000	658000	1781250 / 93000	523500				
7	5-7	King, A.L.	77500		77500					
8	5-7	King, J.E.	77500		67500	10000				
9	5-7	West, D.L.	77500		67500	10000				
10	5-14	White, D.L.	120000		72000	48000				
11	5-14	King, A.L.	80000		40000	40000				
12	5-14	King, J.E.	80000		40000	40000				
13	5-14	West, D.L.	100000		60000	40000				
14	5-21	White, D.L.	120000		48000	72000				
15	5-21	King, A.L.	100000		40000	60000				
16	5-21	King, J.E.	100000		30000	70000				
17	5-21	West, D.L.	90000		50000	40000				
18	5-28	White, D.L.	45000			45000				
19	5-28	King, A.L.	97500		30000	67500				
20	5-28	King, J.E.	97500		37500	60000				
21	5-28	West, D.L.	77500		10000	67500				
22	6-4	White, D.L.	4393750 / 80000	658000 / 66000	2544250 / 24000	1193500				
23	6-4	King, A.L.	75000	55000	20000					
24	6-4	King, J.E.	91250	15000	76250					
25	6-4	West, D.L.	40000			40000				
26	6-11	White, D.L.	45000	4500	40500					
27	6-11	King, A.L.	45000	3750	17500	23750				
28	6-11	King, J.E.	45000	3750	17500	23750				
29	6-18	White, D.L.	76500		76500					
30	6-18	King, A.L.	80000		20000	60000				
		Total (carry forward)	4983500	806000	2836500	1341000				

Consolidated report of labor costs per job
Figure 2-4B

	Date	Name of Employee	1 Gross Wages	2 Baker Job	3 Brown Job	4 Green Job	5	6	7	8
1		(Brought forward)	4983500	8060000	2836500	1341000				
2	6-18	King, J.E.	80000		63750	16250				
3	6-25	White, D.L.	82500		24000	58500				
4	6-25	Lester, C.A.	58500			58500				
5	6-25	King, A.L.	65000		16250	48750				
6	6-25	King, J.E.	48750		20000	28750				
7	6-25	West, D.L.	48750			48750				
8	7-2	White, D.L.	5387000 / 120000	8060000	2960500	1620500 / 120000				
9	7-2	Lester, C.A.	120000			120000				
10	7-2	King, A.L.	100000			100000				
11	7-2	King, J.E.	100000			100000				
12	7-2	West, D.L.	100000			100000				
13	7-9	Lester, C.A.	72000			72000				
14	7-9	King, A.L.	80000			80000				
15	7-9	King, J.E.	80000			80000				
16	7-9	West, D.L.	80000			80000				
17	7-16	King, A.L.	100000			100000				
18	7-16	King, J.E.	100000			100000				
19	7-16	West, D.L.	80000			80000				
20	7-16	Lewis, R.R.	96000			96000				
21	7-23	Lewis, R.R.	96000			96000				
22	7-23	Lester, C.A.	120000			120000				
23	7-23	West, D.L.	100000			100000				
24	7-23	Kidd, L.H.	100000			100000				
25	7-30	Lester, C.A.	120000			120000				
26	7-30	West, D.L.	80000			80000				
27	7-30	Kidd, L.H.	80000			80000				
28	8-6	Lewis, R.R.	7294000 / 120000	8060000	2960500	3524500 / 120000				
29	8-6	Lester, C.A.	120000			120000				
30	8-6	King, A.L.	100000			100000				
		Total (carry forward)	7634000	8060000	2960500	3864500				

Consolidated report of labor costs per job
Figure 2-4C

45

Date	Name of Employee	1 Gross Wages	2 Baker Job	3 Brown Job	4 Green Job	5	6	7	8
1	(Brought forward)	7631000	806000	2960500	3864500				
2	8-6	King, J.E.	100000			100000				
3	8-6	West, D.L.	90000			90000				
4	8-6	Kidd, L.H.	100000			100000				
5	8-13	Lewis, R.R.	24000			24000				
6	8-13	Lester, C.A.	120000			120000				
7	8-13	King, A.L.	100000			100000				
8	8-13	King, J.E.	100000			100000				
9	8-13	West, D.L.	100000			100000				
10	8-20	Lester, C.A.	117000			117000				
11	8-20	King, A.L.	97500			97500				
12	8-20	King, J.E.	97500			97500				
13	8-20	West, D.L.	37500			37500				
14	8-27	Lester, C.A.	120000			120000				
15	8-27	King, A.L.	100000			100000				
16	8-27	King, J.E.	100000			100000				
17	8-27	West, D.L.	80000	806000	2960500	80000 / 5348000				
18	9-3	Lester, C.A.	9114500 / 120000			120000				
19	9-3	King, A.L.	100000			100000				
20	9-3	King, J.E.	100000			100000				
21	9-3	West, D.L.	40000			40000				
22	9-10	Lester, C.A.	96000			96000				
23	9-10	King, A.L.	80000			80000				
24	9-10	King, J.E.	80000			80000				
25	9-10	West, D.L.	80000			80000				
26	9-17	Lester, C.A.	120000			120000				
27	9-17	King, A.L.	100000			100000				
28	9-17	King, J.E.	80000			80000				
29	9-17	West, D.L.	60000			60000				
30	9-24	Lester, C.A.	96000			96000				
	Total (carry forward)	10266500	806000	2960500	6500000				

Consolidated report of labor costs per job
Figure 2-4D

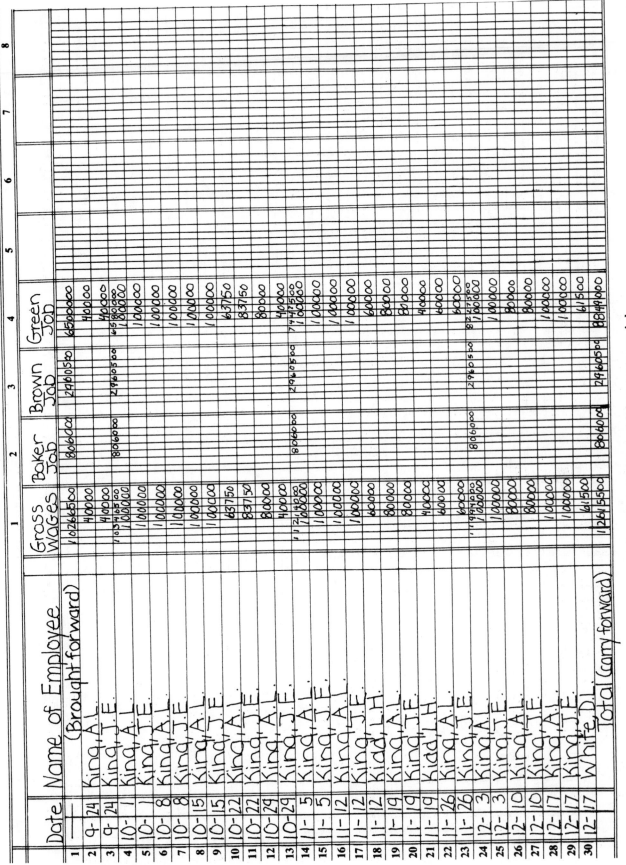

Date	Name of Employee	1 Gross Wages	2 Baker Job	3 Brown Job	4 Green Job	5	6	7	8
	(Brought forward)	10266500	806000	2960550	6500000				
9-24	King, A.L.	40000			40000				
9-24	King, J.E.	40000			40000				
10-1	King, A.L.	1034 8500 / 100000	806000	2960550	6580000 / 100000				
10-1	King, J.E.	100000			100000				
10-8	King, A.L.	100000			100000				
10-8	King, J.E.	100000			100000				
10-15	King, A.L.	100000			100000				
10-15	King, J.E.	100000			63750				
10-22	King, A.L.	63750			83750				
10-22	King, J.E.	83750			80000				
10-29	King, A.L.	80000			40000				
10-29	King, J.E.	40000 / 11212500 / 100000	806000	2960550	7447500 / 100000				
11-5	King, A.L.	100000			100000				
11-5	King, J.E.	100000			100000				
11-12	King, A.L.	100000			100000				
11-12	King, J.E.	100000			60000				
11-12	Kidd, L.H.	60000			80000				
11-19	King, A.L.	80000			80000				
11-19	King, J.E.	80000			40000				
11-19	Kidd, L.H.	40000			60000				
11-26	King, A.L.	60000			60000				
11-26	King, J.E.	60000 / 11994500 / 100000	806000	2960550	8247500 / 100000				
12-3	King, A.L.	100000			100000				
12-3	King, J.E.	80000			80000				
12-10	King, A.L.	80000			80000				
12-10	King, J.E.	80000			100000				
12-17	King, A.L.	100000			100000				
12-17	King, J.E.	100000			61500				
12-17	White, D.L.	61500							
	Total (carry forward)	12615500	806000	2960550	8849000				

Consolidated report of labor costs per job
Figure 2-4E

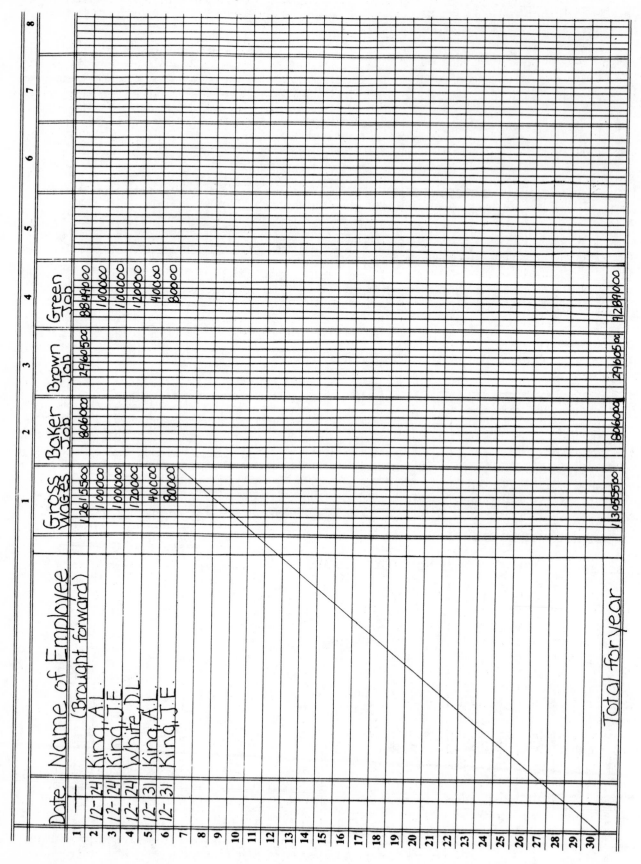

Date	Name of Employee (Brought forward)	1 Gross Wages	2 Baker Job	3 Brown Job	4 Green Job
1		1261550	80600	296050a	884900
2	12-24	King, A.L.	100000			100000
3	12-24	King, J.E.	100000			100000
4	12-24	White, D.L.	120000			120000
5	12-31	King, A.L.	40000			40000
6	12-31	King, J.E.	80000			80000
7						
8						
9						
10						
11						
12						
13						
14						
15						
16						
17						
18						
19						
20						
21						
22						
23						
24						
25						
26						
27						
28						
29						
30	Total for year	1309550	80600	296050a	1289900

Consolidated report of labor costs per job
Figure 2-4F

Here's how to fill out the consolidated report of labor costs per job shown in Figure 2-4A. Use the payroll receipts to enter the following information:

1) Date. The date on this report must be the same as the date shown on the payroll receipts.

2) Name of employee.

3) Total gross wages. Column 1 is for the total gross wages earned by the employee for all jobs he worked during the week.

4) Gross wages per job. Columns 2, 3 and 4 show the portion of gross wages charged to each job. In this example, column 2 is the Baker job, column 3 is the Brown job, and column 4 is the Green job. The total of columns 2, 3 and 4 (on each line) must always equal the amount shown in column 1 (on each line). *If it doesn't, there's an error. Find it and make the correction immediately.*

Look at Figure 2-2A. The payroll receipts for the week ending January 8 show that three employees worked on the Baker job. Enter this information on the consolidated report of labor costs per job shown in Figure 2-4A. In the gross wages column, enter $480.00 for D.L. White, $425.00 for A.L. King, and $425.00 for J.E. King. These employees worked only on the Baker job. Enter their earnings again in column 2.

Notice that a running total has been entered (in pencil) for the gross wages paid from January 8 through February 19. The Baker job was the only job worked during this period. The total labor cost for the Baker job for this period was $3,655.00. The total gross wages for this period were also $3,655.00. We can see at a glance that our entries are correct.

During the week ending March 12, work resumed on the Brown job. Also, survey work began on the Green job. The gross wages shown in column 1 are charged to each of the jobs the employees worked that week.

The total gross wages of all employees shown on this report must equal the total earnings of all employees shown on the individual employee earnings records. For example, on the consolidated report of labor costs per job, the total gross wages paid through the quarter ending March 26 were $12,655.00. On the individual employee earnings records, the total earnings for all employees for the first quarter were:

Arnold L. King	$3,225.00
Jerry E. King	3,025.00
Daniel L. White	6,405.00
Total	$12,655.00

The totals match. Our arithmetic is correct. Now we can prepare our quarterly reports.

Preparing Quarterly Reports

At the end of each quarter, you must file three reports with the federal and state governments: Employer's Federal Tax Return (Form 941); Employer's State Return of Income Tax Withheld; and Employer's Quarterly Contribution Report. Let's look at each of these reports and how to prepare them.

Employer's Quarterly Federal Tax Return (Form 941)

Figure 2-5 shows an Employer's Quarterly Federal Tax Return (Form 941). The individual employee earnings record will give you the information you need to prepare Form 941. This report must include:

1) Number of employees for the quarter. Our individual employee earnings records show that there were three workers employed for the first quarter. Enter this total on line 1.

2) Total wages subject to withholding. These are total wages *before deductions*. The individual employee earnings records show that the total gross earnings for the three employees were $12,655.00 for this quarter:

Arnold L. King	$3,225.00
Jerry E. King	3,025.00
Daniel L. White	6,405.00
Total	$12,655.00

Enter this amount on line 2.

3) Total federal income tax withheld from wages. This is total federal withholding tax for all employees for the quarter:

Arnold L. King	$ 751.30
Jerry E. King	655.70
Daniel L. White	1,517.30
Total	$2,924.30

Enter this amount on line 3.

4) Adjustment of withheld income tax. No adjustments were made. Enter zero on line 4.

5) Adjusted total of income tax withheld. No adjustments of withheld income tax were made for this quarter. Enter on line 5 the same amount as shown on line 3 ($2,924.30).

6) FICA tax. To compute this tax, multiply the taxable FICA wages times the tax rate shown on Form 941. The taxable FICA wages are the same as the

Form **941**

Dept-tment of the Treasury
Internal Revenue Service

Employer's Quarterly Federal Tax Return

Type or print in this space your name, address, and employer identification number as shown on original.

WW 54-0560755

Return for calendar quarter ending
(Enter month and year as on original)

Employer's Name
and
Address

March 31, 19XX

YOUR COPY

1 Number of employees (except household) employed in the pay period that includes March 12th (complete for first quarter only)		3
2 Total wages and tips subject to withholding, plus other compensation . . . . ➤		12,655 00
3 Total income tax withheld from wages, tips, annuities, gambling, etc. (see instructions)		2,924 30
4 Adjustment of withheld income tax for preceding quarters of calendar year		-.0-
5 Adjusted total of income tax withheld . . . . ➤		2,924 30
6 Taxable FICA wages paid . . . . $ 12,655.00 multiplied by 13.4% =TAX .		1,695 77
7 Taxable tips reported . . . . $ .... multiplied by 6.7% =TAX . .		
8 Total FICA taxes (add lines 6 and 7) . . . . ➤		1,695 77
9 Adjustment of FICA taxes (see instructions)		- 0 -
10 Adjusted total of FICA taxes . . . . ➤		1,695 77
11 Total taxes (add lines 5 and 10)		4,620 07
12 Advance earned income credit (EIC) payments, if any (see instructions)		- 0 -
13 Net taxes (subtract line 12 from line 11)		4,620 07

Record of Federal Tax Deposits (See instructions on page 4)

Deposit period ending:	I. Tax liability for period	II. Date of deposit	III. Amount deposited
Overpayment from previous quarter			
First month of quarter — 1st through 7th day			
8th through 15th day			
16th through 22d day			
23d through last day			
A First month total . . . . [A]	528.25	2-15-XX	528.25
Second month of quarter — 1st through 7th day			
8th through 15th day			
16th through 22d day			
23d through last day			
B Second month total . . . . . . [B]	615.92	3-15-XX	615.92
Third month of quarter — 1st through 7th day			
8th through 15th day			
16th through 22d day			
23d through last day			
C Third month total . . . . [C]	3,475.90	4-15-XX	3,475.90
D Total for quarter (add items A, B, and C) .	4,620.07		4,620.07
E Final deposit made for quarter. (Enter zero if the final deposit made for the quarter is included in item D)			- 0 -

14 Total deposits for quarter (including final deposit made for quarter) and overpayment from previous quarter. (See instructions for deposit requirements on page 4) 4,620 07

Note: If undeposited taxes at the end of the quarter are $200 or more, deposit the full amount with an authorized financial institution or a Federal Reserve bank according to the instructions on the back of the Federal Tax Deposit Form 501. Enter this deposit in the Record of Federal Tax Deposits and include it on line 14.

15 Undeposited taxes due (subtract line 14 from line 13—this should be less than $200). Pay to Internal Revenue Service and enter here . . . . ➤ - 0 +

16 If line 14 is more than line 13, enter overpayment here ▶ $ _____ and check if to be: ☐ Applied to next return, or ☐ Refund

17 If you are not liable for returns in the future, write "FINAL" (See instructions) ▶ _____ Date final wages paid ▶

Employer's Quarterly Federal Tax Return (Form 941)
Figure 2-5

| FORM VA 5A EMPLOYER'S RETURN of VIRGINIA INCOME TAX WITHHELD | | | | DO NOT USE THIS SPACE | | |

Employer's state return of income tax withheld
Figure 2-6

total wages shown on line 2 ($12,655.00). In our example, the tax rate is 13.4%. (The employee pays 6.7%, and the employer pays 6.7%.) Our total FICA tax is $1,695.77. Enter this amount on line 6.

7) Taxable tips do not apply. Leave line 7 blank.

8) Total FICA tax due. Same as line 6 ($1,695.77). Enter this amount on line 8.

9) Adjustment of FICA tax. No adjustments were made. Enter zero on line 9.

10) Adjusted total of FICA tax. Same as line 8 ($1,695.77). Enter this amount on line 10.

11) Total taxes. Add the adjusted total income tax from line 5 ($2,924.30) plus the adjusted total FICA tax from line 10 ($1,695.77). Enter the new total ($4,620.07) on line 11.

12) EIC payments, if any. There were no EIC payments. Enter zero on line 12.

13) Net tax. In our example, the net tax due is equal to the income tax withheld from wages ($2,924.30) plus the FICA tax ($1,695.77). The total ($4,620.07) is due and payable either by monthly deposits or to the I.R.S. at the time you file the report. Enter the net tax total on line 13.

14) Total deposits for the quarter. Form 941 shows a record of the federal tax deposits made during the quarter. In our example, $528.25 was deposited on February 15, $615.92 on March 15, and $3,475.90 on April 15. The deposits total $4,620.07. Enter this total on line 14.

15) Undeposited taxes due. Deduct total deposits (line 14) from net tax due (line 13). In our example, the balance is zero. Enter zero on line 15.

We've seen the Employer's Quarterly Federal Tax Return. Now let's look at the Employer's State Return of Income Tax Withheld.

Employer's State Return of Income Tax Withheld
This report may vary from state to state. But all states will request that you show how much state income tax was withheld for each quarter. Again, the individual employee earnings record will give you the figures necessary for this report.

Figure 2-6 shows an Employer's State Return of Income Tax Withheld. Here's how to fill out this report.

On line 1, enter the total state withholding tax for all three employees for the quarter:

VIRGINIA EMPLOYMENT COMMISSION
EMPLOYERS QUARTERLY CONTRIBUTION REPORT

FOR QUARTER ENDING *March 31, 19xx*

Employer's name and address

INSTRUCTIONS ARE ON BACK OF EMPLOYERS COPY OF CONTRIBUTION REPORT

Fed. ID # _54-0560755_
Enter, if missing, change if incorrect

TAX RATE → 0.07% (MULTIPLY BY .0007)

ALSO:
Number of workers on attached payroll by actual count. *3*

If no wages were paid during this quarter write the word "None" on lines 1, 3, 4 & 8, sign and return

PAYROLL DATA

Notice of Change ☐ Name change	1. TOTAL WAGES for quarter, including remuneration other than cash, and including payments over $6,000 per year, per individual.	$ *12,655.00*
☐ Mailing Address change	2. LESS WAGES paid during quarter to each employee in excess of $6,000.	$ *405.00*
☐ Dissolved, no successor	3. WAGES subject to contribution, Line 1 minus Line 2.	$ *12,250.00*

CALCULATION OF CONTRIBUTION

☐ Ownership change, sold or merged with successor ☐ in whole ☐ in part	4. CONTRIBUTION - Multiply total of Line 3 by tax rate shown above.	$ *8.58*
Date of Change	5. CREDIT MEMO NO. ( ) DEDUCT (Always attach white copy of Credit Memos.)	$ *-0-*
	6. INTEREST (computed on contribution - - Line 4 - - at rate of 1% per month from due date to date of payment.)	$ *-0-*
Mo. Day Year	7. PENALTY - $30 for each report filed after the final due date . . . SEE INSTRUCTIONS.	$ *-0-*
Indicate new Name and/or address in this space. If there is a new owner to business indicate new owner's name and address in this space. ➤	8. TOTAL AMOUNT DUE for which remittance is enclosed.	$ *8.58*

CERTIFICATION

I, (or we) certify that the information contained in this report, required in accordance with the Virginia Unemployment Compensation Act, is true and correct and that no part of the contribution reported was, or is to be, deducted from worker's wages.

Employer's Phone Number _____

Bookkeeper's Phone Number _____

Signature *Signed by employer (or his agent)*

Title _____

Date _____

EMPLOYER'S COPY - DETACH AND RETAIN FOR YOUR FILE
VEC FC-20 (R-1-1-79) (250M)

Employer's Quarterly Contribution Report (FUTA)
Figure 2-7A

VIRGINIA EMPLOYMENT COMMISSION
EMPLOYERS QUARTERLY PAYROLL REPORT

FOR QUARTER ENDING **March 31, 19XX**

State ID # 1521-185-041955-0

Employer's name and address

Read carefully instructions on reverse side

Column 1 Worker's Federal Social Security Number	Column 2 Worker's Name First, Middle Initial & Last	Column 3 Total Wages Including Excess over $6,000.00
000-00-0000	1. Arnold L. King	3,225.00
000-00-0000	2. Jerry E. King	3,025.00
000-00-0000	3. Daniel L. White	6,405.00
	4.	
	5.	
	6.	
	7.	
	8.	
	9.	
	10.	
	11.	
	12.	
	13.	
	14.	
	15.	
	16.	
	17.	
	18.	
	19.	
	20.	

Sheet No. **2** of **2** Sheets

Total for this sheet 12,655.00

If more than one sheet of this report, total each sheet and prepare summary sheet

PAYROLL REPORT — DETACH AND RETAIN FOR YOUR FILE

Signed ___Signature of employer___
(or his agent).

Employer's Quarterly Payroll Report
Figure 2-7B

Arnold L. King	$ 162.72
Jerry E. King	148.58
Daniel L. White	318.70
Total	$ 630.00

On lines 2 through 5, enter the appropriate adjustments, penalties, and interest. On line 6, enter the total amount due and payable. This amount is due when you file the return.

Now let's look at the third important quarterly report, the Employer's Quarterly Contribution Report.

Employer's Quarterly Contribution Report (FUTA)

Figure 2-7A shows an Employer's Quarterly Contribution Report (FUTA). Figure 2-7B is an Employer's Quarterly Payroll Report. The individual employee earnings record will provide you with the figures for both reports.

The Employer's Quarterly Payroll Report will vary from state to state. The Employer's Quarterly Contribution Report requires the following information:

1) Total wages. These are total wages subject to withholding for all employees for the quarter. In our example, the total amount is $12,655.00. Enter this amount on line 1.

2) Excess wages. In our example, $6,000.00 per employee per quarter is the maximum wage subject to FUTA tax. Any wages paid in excess of $6,000.00 (per employee per quarter) should be entered on line 2. Look at Figure 2-7B. D.L. White was the only employee whose total wages exceeded the $6,000.00 maximum. Enter the excess ($405.00) on line 2 in Figure 2-7A.

3) Wages subject to contribution. Subtract excess wages (line 2) from total wages (line 1). Enter the new total ($12,250.00) on line 3.

4) Contribution. Multiply the total from line 3 ($12,250.00) times the tax rate shown above line 1 (0.07% or 0.0007) to get the amount of contribution due ($8.58). Enter this amount on line 4.

5) Credits. There were no credits. Enter zero here.

6) Interest. There was no interest. Enter zero here.

7) Penalty. There were no penalty charges. Enter zero here.

8) Total amount due. In our example, the total amount due is $8.58. Enter this amount on line 8.

The total is due and payable when the report is filed.

Be sure to check your arithmetic before filing your quarterly reports. An error on a quarterly report can result in costly corrections when it comes time to file your year-end reports.

Prepare your fourth-quarter quarterly reports before preparing any year-end reports. And remember, *the totals shown on the quarterly reports must balance with the totals shown on your W-2 Forms (total income tax withheld, total wages and total FICA employee tax withheld).*

Preparing Year-end Reports

After you've prepared your quarterly reports, you can do your year-end reports. There are four important year-end reports: Wage and Tax Statements (Form W-2); Transmittal of Income and Tax Statements (Form W-3); Employer's Annual Federal Unemployment Tax Return (Form 940); and Employer's Annual Summary of Income Tax Withheld. The individual employee earnings record has all of the information you need for these reports. Let's look at each of the year-end reports and how to prepare them.

Wage and Tax Statement (Form W-2)

Figure 2-8 shows seven examples of the Wage and Tax Statement (Form W-2). Every Form W-2 must have the employee's name, address and social security number. It must also include the following information taken from the "Total Year" line on the individual employee earnings record:

1) Federal income tax withheld. In Figure 2-8 A, the total federal withholding comes to $8,384.98. Enter this amount in box 11.

2) Wages, tips and other compensation. (This is the same as the total earnings shown on the individual employee earnings record.) In Figure 2-8 A, total wages are $35,225.00. Enter this amount in box 12.

3) FICA tax withheld. (This is the same as social security on the individual employee earnings record.) In Figure 2-8 A, total FICA tax for the year was $2,360.09. Enter this amount in box 13.

4) Total FICA wages (up to the maximum wage subject to FICA tax). For the examples shown in Figure 2-8, we'll use a maximum annual wage of $35,700.00 and a tax rate of 6.7%. In Figure 2-8 A, total FICA wages were $35,225.00. Enter this amount in box 14.

Now look at Figure 2-8 B. Notice that J.E. King's total earnings ($36,187.50) exceeded the maximum wage ($35,700.00) by $487.50. In box 14

1 Control number	222	2 Employer's State number 540560755		
3 Employer's name, address, and ZIP code	4 Sub-total ☐ Cor-rection ☐ Void ☐	Make No Entry Here		
Employer's name and address	7 Employer's identification number 54-0560755			
10 Employee's social security number 000 00 0000	11 Federal income tax withheld $8,384.98	12 Wages, tips, other compensation $35,225.00	13 FICA tax withheld $2,360.09	14 Total FICA wages $35,225.00
15 Employee's name (first, middle, last) Arnold L. King	16 Pension plan coverage? Yes/No No	17	18 FICA tips
Employees address	20 State Income tax withheld $1,791.65	21 State wages, tips, etc. $35,225.00	22 Name of State Virginia
19 Employee's address and ZIP code	23 Local Income tax withheld	24 Local wages, tips, etc.	25 Name of locality
Wage and Tax Statement 19XX		Department of the Treasury-Internal Revenue Service

Form **W-2**

A   Arnold L. King

1 Control number	222	2 Employer's State number 540560755		
3 Employer's name, address, and ZIP code	4 Sub-total ☐ Cor-rection ☐ Void ☐	Make No Entry Here		
Employer's name and address	7 Employer's identification number 54-0560755			
10 Employee's social security number 000 00 0000	11 Federal income tax withheld $8,334.13	12 Wages, tips, other compensation $36,187.50	13 FICA tax withheld $2,391.90	14 Total FICA wages $35,700.00
15 Employee's name (first, middle, last) Jerry E. King	16 Pension plan coverage? Yes/No No	17	18 FICA tips
Employees address	20 State Income tax withheld $1,813.52	21 State wages, tips, etc. $36,187.50	22 Name of State Virginia
19 Employee's address and ZIP code	23 Local Income tax withheld	24 Local wages, tips, etc.	25 Name of locality
Wage and Tax Statement 19XX		Department of the Treasury-Internal Revenue Service

Form **W-2**

B   Jerry E. King

1 Control number	222	2 Employer's State number 540560755		
3 Employer's name, address, and ZIP code	4 Sub-total ☐ Cor-rection ☐ Void ☐	Make No Entry Here		
Employer's name and address	7 Employer's identification number 54-0560755			
10 Employee's social security number 000 00 0000	11 Federal income tax withheld $5,486.90	12 Wages, tips, other compensation $22,050.00	13 FICA tax withheld $1,477.37	14 Total FICA wages $22,050.00
15 Employee's name (first, middle, last) Daniel L. White	16 Pension plan coverage? Yes/No No	17	18 FICA tips
Employee's address	20 State Income tax withheld $1,127.39	21 State wages, tips, etc. $22,050.00	22 Name of State Virginia
19 Employee's address and ZIP code	23 Local Income tax withheld	24 Local wages, tips, etc.	25 Name of locality
Wage and Tax Statement 19XX		Department of the Treasury-Internal Revenue Service

Form **W-2**

C   Daniel L. White

Wage and tax statement (Form W-2)
Figure 2-8

1 Control number		2 Employer's State number			
	222	540560755			

3 Employer's name, address, and ZIP code

Employer's name and address

4 Sub-total ☐	Cor-rection ☐	Void ☐

7 Employer's identification number
54-0560755

Make No Entry Here

10 Employee's social security number	11 Federal income tax withheld	12 Wages, tips, other compensation	13 FICA tax withheld	14 Total FICA wages
000 00 0000	$4,346.85	$15,937.50	$1,067.83	$15,937.50

15 Employee's name (first, middle, last)	16 Pension plan coverage? Yes/No	17	18 FICA tips
Dennis L. West	NO		

Employee's address

20 State Income tax withheld	21 State wages, tips, etc.	22 Name of State
$811.07	$15,937.50	Virginia
23 Local Income tax withheld	24 Local wages, tips, etc.	25 Name of locality

19 Employee's address and ZIP code

Wage and Tax Statement 19 XX

Form **W-2**   Department of the Treasury-Internal Revenue Service

D Dennis L. West

1 Control number		2 Employer's State number			
	222	540560755			

3 Employer's name, address, and ZIP code

Employer's name and address

4 Sub-total ☐	Cor-rection ☐	Void ☐

7 Employer's identification number
54-0560755

Make No Entry Here

10 Employee's social security number	11 Federal income tax withheld	12 Wages, tips, other compensation	13 FICA tax withheld	14 Total FICA wages
000 00 0000	$3,490.20	$13,995.00	$937.67	$13,995.00

15 Employee's name (first, middle, last)	16 Pension plan coverage? Yes/No	17	18 FICA tips
Clarence A. Lester	No		

Employee's address

20 State Income tax withheld	21 State wages, tips, etc.	22 Name of State
$712.66	$13,995.00	Virginia
23 Local Income tax withheld	24 Local wages, tips, etc.	25 Name of locality

19 Employee's address and ZIP code

Wage and Tax Statement 19 XX

Form **W-2**   Department of the Treasury-Internal Revenue Service

E Clarence A. Lester

1 Control number		2 Employer's State number			
	222	540560755			

3 Employer's name, address, and ZIP code

Employer's name and address

4 Sub-total ☐	Cor-rection ☐	Void ☐

7 Employer's identification number
54-0560755

Make No Entry Here

10 Employee's social security number	11 Federal income tax withheld	12 Wages, tips, other compensation	13 FICA tax withheld	14 Total FICA wages
000 00 0000	$806.70	$3,360.00	$225.12	$3,360.00

15 Employee's name (first, middle, last)	16 Pension plan coverage? Yes/No	17	18 FICA tips
Robert R. Lewis	No		

Employee's address

20 State Income tax withheld	21 State wages, tips, etc.	22 Name of State
$168.27	$3,360.00	Virginia
23 Local Income tax withheld	24 Local wages, tips, etc.	25 Name of locality

19 Employee's address and ZIP code

Wage and Tax Statement 19 XX

Form **W-2**   Department of the Treasury-Internal Revenue Service

F Robert R. Lewis

Wage and tax statement (Form W-2)
Figure 2-8 (continued)

1 Control number		2 Employer's State number				
	222	540560755				
3 Employer's name, address, and ZIP code		4 Sub-total	Cor-rection	Void		
Employer's name and address		☐	☐	☐	Make No Entry Here	
		7 Employer's identification number				
		54-0560755				
10 Employee's social security number	11 Federal income tax withheld	12 Wages, tips, other compensation		13 FICA tax withheld		14 Total FICA wages
000 00 0000	$1,005.50	$3,800.00		$254.60		$3,800.00
15 Employee's name (first, middle, last)		16 Pension plan coverage? Yes/No		17		18 FICA tips
Lawson H. Kidd		No				
		20 State income tax withheld		21 State wages, tips, etc.		22 Name of State
Employee's address		$190.32		$3,800.00		Virginia
		23 Local income tax withheld		24 Local wages, tips, etc.		25 Name of locality
19 Employee's address and ZIP code						
Wage and Tax Statement		19 XX				

Form **W-2**　　　　　　　　　　　　　　　　　　Department of the Treasury-Internal Revenue Service

G Lawson H. Kidd

Wage and tax statement (Form W-2)
Figure 2-8 (continued)

for J.E. King, enter only the wages subject to FICA tax ($35,700.00). And in box 13 for J.E. King, you must enter only the tax on $35,700.00. To find the correct tax, multiply $35,700.00 times 0.067. The correct tax is $2,391.90.

Look at the individual employee earnings record for J.E. King. Notice that the total social security tax withheld from his wages was $2,424.58, not $2,391.90. This happened because J.E. King's total wages didn't exceed the maximum until the last pay period of the year. You must pay him the difference between the amount withheld ($2,424.58) and the actual amount due ($2,391.90). This comes to $32.68.

Be sure to correct the numbers on your quarterly reports so that they match the corrected totals on your year-end reports.

5) State income tax withheld. (This is the same as state withholding tax on the individual employee earnings record.) In our example in Figure 2-8 A, total state income tax came to $1,791.65. Enter this amount in box 20.

6) State wages, tips, etc. (This is the same as total earnings on the individual employee earnings record.) In our example, total earnings for the year were $35,225.00. Enter this amount in box 21.

Transmittal of Income and Tax Statements (Form W-3)
Figure 2-9 shows a Transmittal of Income and Tax Statement (Form W-3). File this report with the W-2 Forms that you send to the federal government at the end of the year.

Form W-3 must include the following information:

1) Number of attached Form W-2 statements. Enter this number in box 16. In this example, there are seven Form W-2 statements.

2) Date of report. Enter the date in box 17.

3) Total FICA tax withheld. This is the total of all FICA taxes reported in box 13 on the W-2 Forms you're enclosing with this Form W-3. In this example, the total is $8,714.58. Enter this amount in box 18 of Form W-3.

4) Total federal income tax withheld. This is the total of all federal income taxes reported in box 11 on the W-2 Forms you're enclosing with this Form W-3. In our example, it's $31,855.26. Enter this amount in box 19 of Form W-3.

5) Total FICA wages. This is the total of all FICA wages reported in box 14 on the enclosed W-2 Forms. In our example, it comes to $130,067.50. Enter this amount in box 21 of Form W-3.

6) Total wages, tips, and other compensation. This is the total of all wages reported in box 12 on the enclosed W-2 Forms. In our example, this comes to $130,555.00. Enter this amount in box 22 of Form W-3.

1 Control number		2 Employee's State number					
	ӠӠӠ	540560755					
Kind of Tax Statements Transmitted ▷	3 Official use ☐	4 Military ☐	5 Agriculture ☐	6 W-2 ☒	7 Original ☒	8 With TIN ☒	
	9 Railroad ☐	10 Household ☐	11 State or Local Gov. ☐	12 W-2P ☐	13 Corrected ☐	14 Without TIN ☐	
15 State SSA number		16 Number of statements attached 7				17 Date of report 01-03-19XX	
18 Total FICA tax withheld $8,714.58		19 Total Federal income tax withheld $31,855.26				20 Total FICA tips	
21 Total FICA wages $130,067.50		22 Total wages, tips, and other compensation $130,555.00				23 Gross annuities, pensions, retired pay, etc.	
24 Employer's identification number 54-0560755				25 Establishment number		26 Taxable annuities, pensions, retired pay, etc.	
Employer's Name and Address 27 Employer's name, address and ZIP code			Internal Revenue Service—Department of the Treasury Form **W-3** Transmittal of Income and Tax Statements 19 XX				

Transmittal of income and tax statement (Form W-3)
Figure 2-9

Employer's Annual Federal Unemployment Tax Return (Form 940)

This is the federal FUTA tax the employer must pay on all employees. Figure 2-10 shows an Employer's Annual Federal Unemployment Tax Return (Form 940). This report requires the following information:

1) Name of state to which you pay your contribution. In our example, the state is Virginia (VA). Enter this information in column 1.

2) State reporting number. In our example, this number is 1521-185-041955-0. Enter this number in column 2. (This number is also shown at the top of your Employer's Quarterly Contribution Report. See Figure 2-7A.)

3) Taxable payroll. This is equal to the total annual gross wages shown on your consolidated report of labor costs per job. (See Figure 2-4F.) In our example, the total gross wages for the year come to $130,555.00. Enter this amount in column 3. This amount also appears in box 22 on the Transmittal of Income and Tax Statement (Form W-3) shown in Figure 2-9.

4) Experience rate period. This is the period of time covered in this report. In our example, the experience rate period begins on January 1 and ends on December 31. Enter this information in column 4.

5) Experience rate. This is the rate granted by the state for the period covered in column 4. In this example, it's 0.07%. Enter this rate in column 5.

6) Contributions had rate been 2.7%. This assumes that the experience rate is 2.7% instead of 0.07%. Multiply column 3 times 2.7%. In our example, this comes to $3,524.99. Enter this amount in column 6. This is where we begin our computation of the federal FUTA tax.

7) Contributions payable. Multiply the payroll shown in column 3 ($130,555.00) times the experience rate shown in column 5 (0.07% or 0.0007). This comes to $91.39. Enter this amount in column 7.

8) Additional credit. Subtract the amount in column 7 ($91.39) from the amount in column 6 ($3,524.99). The result is $3,433.60. Enter this amount in column 8.

9) Contributions actually paid to state. Total the contributions shown on your Employer's Quarterly Contribution Reports. For our example, the annual total comes to $30.80. Enter the annual total in column 9.

10) Total tentative credit. Add column 8 ($3,433.60) plus column 9 ($30.80). This comes to $3,464.40. Enter this amount on line 10.

Form **940**
Department of the Treasury
Internal Revenue Service

Employer's Annual Federal Unemployment Tax Return

19 X X

EMPLOYER'S I.D. NUMBER: W.W. 540560755

EMPLOYER'S NAME AND ADDRESS

Name of State [1]	State reporting number as shown on employer's State contribution returns [2]	Taxable payroll (as defined in State set) [3]	Experience rate period [4] From—	To—	Experience rate [5]	Contributions had rate been 2.7% (col. 3 x 2.7%) [6]	Contributions payable at experience rate (col. 3 x col. 5) [7]	Additional credit (col. 6 minus col. 7) [8]	Contributions actually paid to State [9]
VA	1521-185-041955-0	$130,555.00	1/1/xx	12/31/xx	.07%	3,524.99	91.39	3,433.60	30.80
	Totals ▶	130,555.00						3,433.60	30.80

10 Total tentative credit (column 8 plus column 9— see Instructions on page 4) 3,464.40
11 Total remuneration (including exempt remuneration) paid during the calendar year for services of employees 130,555.00

Exempt Remuneration (See Instructions on Page 4)	(a) New jobs credit wages	(b) Amount paid
12 Exempt remuneration. (Explain each exemption shown, attaching additional sheet if necessary) ▶		
13 Remuneration in excess of the first $4,200 in column (a), and the first $6,000 in column (b), paid to individual employees exclusive of exempt amounts entered on line 12. Do not use State wage limitation ...		86,560.00
14 Total exempt remuneration (line 12 plus line 13)		86,560.00

15 a New jobs credit total wages (subtract line 12, column (a) from line 11 ...
   b New jobs credit wages (subtract line 14, column (a) from line 11) ...
   c Total taxable FUTA wages (subtract line 14, column (b) from line 11) 43,995.00
16 Gross Federal tax (multiply line 15c by .034) 1,495.83
17 Maximum credit (multiply line 15c by .027) 1,187.87
18 Line 10 or line 17 whichever is smaller 1,187.87
19 Amount, if any, of wages on line 15c attributable to Rhode Island $_____ x .003 .... -0-
20 Credit allowable (subtract line 19 from line 18) 1,187.87
21 Net Federal tax (subtract line 20 from line 16) 307.96

Record of Federal Tax Deposits for Unemployment Tax (Form 508)			
Quarter	Liability by period	Date of deposit	Amount of deposit
First	8.58	—	—
Second	17.21	—	—
Third	4.31	—	—
Fourth	277.86	1-4-xx	307.96

22 Total Federal tax deposited 307.96
23 Balance due (subtract line 22 from line 21 —) Pay to Internal Revenue Service .... -0-
24 Overpayment (subtract line 21 from line 22) .... -0-
25 In no longer in business at end of year, write "Final" here .... ▶

1-04-XX   SIGNATURE   TITLE

Employer's annual federal unemployment tax return (Form 940)
Figure 2-10

```
Form VA - 6A
REV. 1-77

EMPLOYER'S ANNUAL OR FINAL SUMMARY OF VIRGINIA INCOME TAX WITHHELD
```

			VA. INCOME TAX PAID AS SHOWN ON FORM VA-5 (s)			
SEE REVERSE SIDE FOR INSTRUCTIONS			1. JAN.		7. JULY	
CALENDAR YEAR	FILE BY THIS DATE	VA. EMPLOYER ACCOUNT NUMBER	2. FEB.		8. AUG.	
19xx	04/30/xx	540560755	3. MAR. 1st Qtr.	630 00	9. SEPT. 3rd Qtr.	2,525 97
			4. APR.		10. OCT.	
Employer's name and address			5. MAY.		11. NOV.	
			6. JUNE 2nd Qtr.	2,092 78	12. DEC. 4th Qtr.	1,366 13
			15. ENTER THE TOTAL NUMBER OF W-2 (STATE COPY) SENT WITH THIS REPORT ___7___		13. TOTAL PAYMENTS (Lines 1 thru 12)	6,614 88
FOR TAXPAYER ASSISTANCE CALL					14. TOTAL VA. TAX WITHHELD ON W-2	6,614 88
					NOTE: IF LINES 13 AND 14 DO NOT AGREE ATTACH EXPLANATION OF THE DIFFERENCE.	

RECC/NP-28454

Employer's annual summary of state income tax withheld
Figure 2-11

11) Total remuneration. In this example, total remuneration is the same amount as shown in column 3 ($130,555.00). Enter this amount on line 11.

12) Excess remuneration. This is the total of all wages paid in excess of the maximum wage subject to FUTA tax. This information is taken from line 2 on the Employer's Quarterly Contribution Report. (See Figure 2-7A.) Add together the excess wages from all four quarters. Enter this amount on line 13 in Figure 2-10.

13) Total taxable FUTA wages. Subtract total exempt remuneration shown on line 14 ($86,560.00) from the total remuneration shown on line 11 ($130,555.00). The result is $43,995.00. Enter this amount on line 15C.

14) Gross federal tax. Multiply total taxable FUTA wages shown on line 15C ($43,995.00) times 0.034. This comes to $1,495.83. Enter this amount on line 16.

15) Maximum credit. Multiply total taxable FUTA wages shown on line 15C ($43,995.00) times 0.027. This comes to $1,187.87. Enter this amount on line 17.

16) Net federal tax. Subtract credit allowable shown on line 20 ($1,187.87) from gross federal tax shown on line 16 ($1,495.83). The result is $307.96. Enter this amount on line 21.

17) Record of federal tax deposits for unemployment tax. List the liability, date of deposit and amount of deposit. Enter on line 22 the total amount of tax deposited. In our example, this comes to $307.96.

18) Balance due. Subtract the total deposit shown on line 22 ($307.96) from the net federal tax shown on line 21 ($307.96). If there's a balance due, enter it on line 23. In our example, there is no balance due.

19) Overpayment. Subtract line 21 from line 22. If there was an overpayment, enter the amount on line 24.

Employer's Annual Summary of State Income Tax Withheld

Figure 2-11 shows an Employer's Annual Summary of State Income Tax Withheld. This report shows the total state income tax withheld from all employees for the year. Prepare this report after all quarterly reports are complete. This report must agree with the combined totals of the state income taxes withheld for each quarter.

The individual employee earnings record will provide you with the information you need for the Employer's Annual Summary of State Income Tax Withheld. This report requires the following information:

1) State income tax withheld and paid for each quarter. Our sample report shows that $630.00 was withheld for the first quarter, $2,092.78 for the second quarter, $2,525.97 for the third quarter, and $1,366.13 for the fourth quarter. Enter each amount in the appropriate space on lines 3, 6, 9 and 12. To cross-check these amounts, add up the quarterly totals on the individual employee earnings records shown in Figures 2-3A through 2-3G.

2) Total payments. Add up the payments for all four quarters. In our example, the total payments come to $6,614.88. Enter this amount on line 13.

3) Total state tax withheld. This is the total of all state income tax withheld for all employees for the year. This information comes from box 20 on your enclosed W-2 Forms. Figure 2-11 shows that our annual total comes to $6,614.88. Enter this amount on line 14. *If the amount shown on line 13 and the amount shown on line 14 are not the same, you must explain the difference. Attach this explanation to your report when you file it.*

4) Total number of W-2 Forms enclosed. These are the *state* copies of the W-2 Forms. In our example, there are seven W-2 Forms enclosed. Enter this number on line 15.

Can you see why it's so important to verify the arithmetic on the individual employee earnings record *before* filing your quarterly reports? If you've made errors, it's easier to correct them over a three-month period. If errors aren't detected until year-end, corrections will cost you a lot of time and money.

In this chapter, we've seen how a simple, efficient payroll record system can simplify your quarterly and year-end reports. In the next chapter, we'll look at an efficient system for billing, income and disbursements, and subcontracted work.

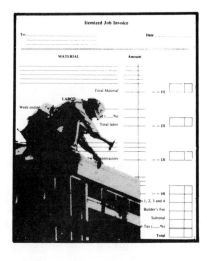

Chapter 3

Billing, Income and Disbursements, and Subcontracted Work

In Chapter 2, we set up a payroll record system that will help you control your labor costs on every job. In this chapter, we'll take a look at the three standard payment agreements for construction work. Then we'll show you a simple, efficient system for handling your billing, income and disbursements, and subcontracted work. We'll also look at how this new system will help you prepare Schedule C (Form 1040) for your income taxes.

Standard Payment Agreements

There are three standard payment agreements for most construction projects: contract, cost-plus percentage, and cost-plus fixed fee. Here's how they work.

Contract

When you build under contract, you agree to build according to plans and for a specified sum of money. But few projects are ever built exactly as specified in the plans. With a contract arrangement, you must try to anticipate unknown circumstances that could increase costs, such as rock or unstable soil. When changes are necessary, you must get a signed agreement from the owner. This agreement should state the additional work re-

quired and the exact cost.

Cost-plus Percentage

If you are building an expensive home, cost-plus percentage is the best agreement you can have. Few owners know (before construction begins) everything they'll want in their finished home. If you're building under contract, it's expensive and time-consuming to prepare cost estimates and get a signed agreement for every change the owner makes. On a cost-plus job, the owner agrees to pay all costs plus a specified percentage of these costs as your fee. If he makes changes along the way, costs increase, and your compensation increases proportionately.

Cost-plus Fixed Fee

This agreement is similar to the cost-plus percentage agreement except that the owner pays all costs plus a fixed fee per hour for your services.

The owner may provide separate agreements for subcontracted work. In this case, the cost-plus fixed fee agreement is to your advantage. It guarantees that you'll be compensated for the time spent coordinating the subcontractors' work on the job. If you had a cost-plus percentage agreement, you wouldn't be paid for this additional time.

Billing
When you enter into a payment agreement with an owner, the agreement should specify when and how you'll be paid. Payments usually come due at the completion of each stage of construction. For example, when the foundation is complete and ready for the subfloor, the first payment is due. When the house is under roof, the second payment is due. When the drywall is finished, the third payment is due. When the house is finished or occupied (whichever comes first), the final payment is due.

When you bill the owner for contract payments, no itemizing is required. Just send a simple statement. But when a house is built on a cost-plus percentage or cost-plus fixed fee basis, each statement sent to the owner should be itemized. Use the itemized job invoice for this purpose.

Itemized Job Invoice
Figure 3-1A shows an itemized job invoice. The date on the invoice should be the same as the date of the last payroll period. For example, if the last payroll period (for labor included on the invoice) is January 29, the date on the itemized job invoice should also be January 29.

The itemized job invoice should include the following information:

1) Materials— Attach a copy of each invoice from every supplier. Total the invoices, and enter the total on the "Total Material" line. Make a notation that you've attached the invoices, as shown in Figure 3-lA.

2) Labor— Attach copies of all weekly time sheets covering work included in the invoice. Add the appropriate percentage for payroll taxes, insurance and overhead. This percentage should be agreed upon with the owner *before construction begins*. Enter the percentage as shown in Figure 3-1B.

3) Subcontracted work— Attach copies of all invoices from subcontractors. Total the invoices, and enter the total as shown in Figure 3-1A.

4) Other charges— Include in this space any charges not included elsewhere.

5) Builder's fee— This is your fee, as agreed upon with the owner.

6) Business and occupation tax (B & O tax)— In some states and localities, this tax is levied on the gross amount received. The percentage of B & O tax should always be entered in the space allotted to it, as shown in Figure 3-1A.

7) Total— This is the amount charged to the owner.

Figures 3-1C through 3-1H show additional examples of itemized job invoices. Notice that the invoices vary with each stage of construction. Some invoices will include materials but no labor or subcontracted work. Other invoices will show only a builder's fee that has accumulated over a period of time. Some construction projects will be located in areas that don't require the B & O tax.

After you've completed and mailed your billing, you'll begin to receive payments from the owners. You'll enter these payments on your income and disbursements record. Let's take a look at this important form.

Income and Disbursements
Figure 3-2A is a ten-column analysis sheet. Use this form for business income and disbursements. This record will help you at tax time. It shows an itemized listing of income and expenses you can refer to when preparing your Schedule C (Form 1040) for your income taxes.

The income and disbursements record should include the following information:

1) Income— Column 1 shows gross receipts. Enter payments received from the owners (Baker, Brown and Green). Remember to record the check number in the column preceding column 1.

2) Disbursements— Column 2 shows money paid out. The disbursement shown in column 2 must also be charged to the correct account in columns 3 through 10. See Figure 3-2A. And column 2 (on each line) must equal the total of columns 3 through 10 (on each line). If they do not balance, there's an error. Find it and correct it immediately. Record all disbursement check numbers in the column preceding column 1.

Itemized Job Invoice

To: *Mr. B. W. Baker*   **Date** *1-3-XX*

MATERIAL	Amount		
Attached invoices	*2926.19*		
Total Material	*2,926.19*	— — [1]	*2926 | 19*

LABOR

	Amount		
Week ending:	*—0—*		
Payroll taxes, insurance and overhead (___%)			
Total Labor	*—0—*	— — [2]	*—0—*

SUBCONTRACTORS

	Amount		
Labor to pour concrete on porch (Invoice attached)	*297 00*		
Total Subcontractors	*297 00*	— — [3]	*297 | 00*

OTHER CHARGES

	Amount		
	—0—		
Total Other Charges	*—0—*	— — [4]	*—0—*
Total of lines 1, 2, 3 and 4			*3,223 | 19*
Builder's Fee			*1,500 | 00*
Subtotal			*4723 | 19*
Business and Occupation Tax (*3.85*%)			*181 | 84*
Total			*4905.03*

Itemized Job Invoice
Figure 3-1A

Itemized Job Invoice

To: _Mr. B. W. Baker_   Date _1-29-XX_

MATERIAL	Amount		
Attached Invoices	723 28		
Total Material 723 28		— — [1]	723 28
LABOR			
Week ending: _Attached payrolls (1-8-XX thru 1-29-XX)_	1975 00		
Payroll taxes, insurance and overhead (25%)	493 75		
Total Labor 2468 75		— — [2]	2468 75
SUBCONTRACTORS			
	—0—		
Total Subcontractors		— — [3]	—0—
OTHER CHARGES			
	—0—		
Total Other Charges		— — [4]	—0—
Total of lines 1, 2, 3 and 4			3,192 03
Builder's Fee			320 00
Subtotal			3512 03
Business and Occupation Tax (3.85%)			135 21
Total			3647 24

Itemized Job Invoice
Figure 3-1B

Itemized Job Invoice

To: *Mr. L. C. Brown*   **Date** *2-12-XX*

MATERIAL	Amount			
Attached invoices	*1421 23*			
Total Material	*1421 23*	– – [1]	*1421*	*23*
LABOR				
Week ending:	*–0–*			
Payroll taxes, insurance and overhead (____%)				
Total Labor	*–0–*	– – [2]	*–0–*	
SUBCONTRACTORS				
	–0–			
Total Subcontractors	*–0–*	– – [3]	*–0–*	
OTHER CHARGES				
	–0–			
Total Other Charges	*–0–*	– – [4]	*–0–*	
Total of lines 1, 2, 3 and 4			*1421*	*23*
Builder's Fee			*–0–*	
Subtotal			*1421*	*23*
Business and Occupation Tax (*N/A* %)			*–0–*	
Total			*1421*	*23*

Itemized Job Invoice
Figure 3-1C

Itemized Job Invoice

To: *Mr. B. W. Baker*   Date *2-26-XX*

MATERIAL   **Amount**

Total Material	-0- — — [1]

	-0-

LABOR

Week ending: _____

Payroll taxes, insurance and overhead (____%)

Total Labor	-0- — — [2]

	-0-

SUBCONTRACTORS

Total Subcontractors	-0- — — [3]

	-0-

OTHER CHARGES

Total Other Charges	-0- — — [4]

	-0-	
Total of lines 1, 2, 3 and 4	-0-	
Builder's Fee	3,000	00
Subtotal	3,000	00
Business and Occupation Tax (*3.85*%)	115	50
Total	3,115	50

Itemized Job Invoice
Figure 3-1D

Itemized Job Invoice

To: *Mr. E. J. Green*   Date *4-23-XX*

MATERIAL	Amount		
Attached invoice	*807 98*		
Total Material	*807 98* — — [1]	*807*	*98*

LABOR			
Week ending: *3-12-XX thru 4-23-XX (Payroll Attached)*	*4,595 00*		
Payroll taxes, insurance and overhead (*25* %)	*1,148 75*		
Total Labor	*5,743 75* — — [2]	*5743*	*75*

SUBCONTRACTORS			
Total Subcontractors	*—0—* — — [3]	*—0*	*—*

OTHER CHARGES			
Fill dirt and dozer (Invoice Attached)	*2751 30*		
Total Other Charges	*2751 30* — — [4]	*2751*	*30*
Total of lines 1, 2, 3 and 4		*9303*	*03*
Builder's Fee		*930*	*00*
Subtotal		*10,233*	*03*
Business and Occupation Tax (*N/A* %)		*—0*	*—*
Total		*10,233*	*03*

Itemized Job Invoice
Figure 3-1E

69

Itemized Job Invoice

To: *Mr. B. W. Baker*   Date *5-4-XX*

MATERIAL	Amount		
Attached Invoices	*1758 21*		
Total Material *1758 21*	— — [1]		*1,758 21*

LABOR

	Amount		
Week ending: *4-2-XX thru 4-30-XX (Payrolls Attached)*	*2925 00*		
Payroll taxes, insurance and overhead (*25*%)	*731 25*		
Total Labor *3656 25*	— — [2]		*3,656 25*

SUBCONTRACTORS

	Amount		
W.C. Martin	*534 33*		
Total Subcontractors *534 33*	— — [3]		*534 33*

OTHER CHARGES

	Amount		
Total Other Charges *—0—*	— — [4]		*—0—*

Total of lines 1, 2, 3 and 4	*5,948 79*
Builder's Fee	*600 00*
Subtotal	*6,548 79*
Business and Occupation Tax (*3.85*%)	*252 13*
Total	*6,800 92*

Itemized Job Invoice
Figure 3-1F

Itemized Job Invoice

To: *Mr. E. J. Green*   Date *6-25-XX*

MATERIAL	Amount			
House package (factory-built home)	46,884 43			
Stone for temporary drive (invoice attached)	143 28			
Total Material	47,027 71	— — [1]	47,027 71	

LABOR

	Amount			
Week ending: 6-18-XX and 6-25-XX (Payrolls attached)	3,395 00			
Payroll taxes, insurance and overhead (25 %)	848 75			
Total Labor	4,243 75	— — [2]	4,243 75	

SUBCONTRACTORS

	Amount			
Total Subcontractors	—0—	— — [3]	—0—	

OTHER CHARGES

	Amount			
Attached invoice for fill dirt	4,397 50			
Total Other Charges	4,397 50	— — [4]	4,397 50	
Total of lines 1, 2, 3 and 4			55,668 96	
Builder's Fee			5,500 00	
Subtotal			61,168 96	
Business and Occupation Tax (N/A %)			—0—	
Total			61,168 96	

Itemized Job Invoice
Figure 3-1G

Itemized Job Invoice

To: _Mr. E. J. Green_   Date _12-31-XX_

MATERIAL	Amount		
Attached invoices	821 69		
Total Material	821 69	— — [1]	821 69

LABOR

Week ending: _12-17-XX thru 12-31-XX (Payrolls attached)_	7,015 00		
Payroll taxes, insurance and overhead (_25_%)	1,753 75		
Total Labor	8,768 75	— — [2]	8,768 75

SUBCONTRACTORS

Total Subcontractors	—0—	— — [3]	—0—

OTHER CHARGES

Total Other Charges	—0—	— — [4]	—0—
Total of lines 1, 2, 3 and 4			9,590 44
Builder's Fee			950 00
Subtotal			10,540 44
Business and Occupation Tax (_N/A_%)			—0—
Total			10,540 44

Itemized Job Invoice
Figure 3-1H

3) Interest— This is shown in Column 3. If you maintain an office in your home, the interest on your home mortgage can be prorated and charged to your business.

4) Taxes— Column 4 shows business taxes. These include your FICA, FUTA, B & O taxes and your contractor's license fee. If you have an office in your home, you can also prorate your real estate taxes according to the amount of space your office occupies. You can charge the prorated portion of the taxes to your business.

5) Office— If you rent your office space, you can enter the rent in column 5. Or you can set up a separate column just for rent. If you have an office in your home, include in column 5 the prorated expenses of heat, lights, telephone, water, and sewer for the house.

6) Vehicle— Vehicle expenses are shown in column 6. Be sure to record the mileage reading from your vehicle's odometer at the beginning of each tax year and again at the end of the year. This gives you an accurate record of the actual miles driven for the year. A good place to record these readings is on the first and last sheets of the income and disbursements record. See Figures 3-2A and 3-2G.

7) Insurance— Enter worker's compensation, liability and vehicle insurance in column 7. If you have an office in your home, you can prorate the insurance on your home and charge the prorated share to your business.

8) Materials and supplies— Columns 8, 9 and 10 show disbursements made for materials and supplies. Be sure to set up a separate column for each job. And make sure that the material ordered for each job is charged to that job. Keep the material delivery tickets in your file until the corresponding invoices are received. This will give you a complete picture of your actual costs plus your outstanding costs-to-date.

On the income and disbursements record, the running totals (in pencil) at the end of each month provide a constant check on the costs-to-date for each job. Annual totals are shown on the last line of Figure 3-2G.

Figures 3-2B through 3-2G show additional examples of income and disbursement records. These records span a period of one year and cover three different jobs: the Baker, Brown and Green jobs.

Subcontracted Work

Figure 3-3 is a record of disbursements to subcontractors. The number of jobs will dictate the number of columns you need. This form will give you an up-to-date picture of each segment of your subcontracted work for each job.

Column 1 is for the total disbursement paid to each subcontractor. This disbursement is then charged to the appropriate jobs in columns 2, 3 and 4. The total shown in column 1 (on each line) must always equal the total of columns 2, 3 and 4 (on each line). The running totals for column 1 must also equal the combined running totals of columns 2, 3 and 4. Look at the running totals (through June) in Figure 3-3:

Baker job	$1,316.33
Brown job	1,840.32
Green job	2,780.00
Total	$5,936.65

This total ($5,936.65) matches the June running total shown in column 1. Our arithmetic is correct.

Figure 3-3 allows you to see the cost breakdown for your subcontracted work on any job at any given time. The breakdown for the Green job through September 30 looks like this:

Masonry (June 2)	$2,780.00
Roofing (August 12)	544.30
Masonry (September 23)	919.69
Masonry (September 30)	683.85
Total	$4,927.84

Figure 3-3 completes our simple, efficient system for handling billing, income and disbursements, and subcontracted work. Now let's look at how this new system can help you prepare Schedule C (Form 1040) for your income taxes.

ODOMETER READING 1-1-XX: 32,363.4

	Date	RECEIPTS AND DISBURSEMENTS	Ck. No	Income (1)	Disbursement (2)	Interest (3)
1	1-5	Postage Stamps	—		20 00	
2	1-5	18.2 Gals. Gasoline	—		23 28	
3	1-7	B. W. Baker	888	4905 03		
4	1-11	9.6 Gals. Gasoline	—		12 28	
5	1-14	10.9 Gals. Gasoline	—		13 94	
6	1-15	Long Distance Call	768		80	
7	1-17	F.I.C.A. (Previous Quarter)	771		702 04	
8	1-17	Federal F.U.T.A (Previous Year)	772		133 39	
9	1-17	State F.U.T.A. (Previous Quarter)	773		310 57	
10	1-19	12.9 Gals. Gasoline	—		16 50	
11	1-21	12.8 Gals. Gasoline	—		16 37	
12	1-26	10.9 Gals Gasoline	—		13 94	
13	1-29	9.1 Gals. Gasoline	—		11 64	
14	2-3	12.8 Gals. Gasoline	—	4905 03	1274 75 / 16 37	—
15	2-5	B. W. Baker	892	3647 24		
16	2-7	January Invoices	797		723 28	
17	2-7	8.3 Gals. Gasoline	—		10 62	
18	2-10	6.3 Gals. Gasoline and Car Wash	—		11 06	
19	2-12	Long Distance telephone calls	799		2 44	
20	2-14	12.8 Gals. Gasoline	—		16 37	
21	2-16	B. W. Baker	894	2275 63		
22	2-18	12.7 Gals. Gasoline	—		16 24	
23	2-19	8.7 Gals. Gasoline	—		11 13	
24	2-23	Office Supplies	—		6 50	
25	2-24	12.8 Gals. Gasoline	—		16 37	
26	2-26	Car Wash	—		3 00	
27	2-26	B. W. Baker	900	1749 87		
28	2-28	February Invoices	815		1091 27	
29	3-1	12.8 Gals. Gasoline	—	12577 77	3199 40 / 16 37	—
30	3-5	Insurance on Vehicle	—		64 91	
		TOTAL (CARRY FORWARD)	—	12577 77	3280 68	—

Income and Disbursements

4	5	6	7	8	9	10
Taxes	Office	Vehicle	Insurance	Baker Job	Brown Job	
	2000					
		2328				
		1228				
		1394				
				80		
70204						
13339						
31057						
		1650				
		1637				
		1394				
		1164				
114600	2000	10795	—	80	—	
		1637				
				72328		
		1062				
		1106				
				98	146	
		1637				
		1624				
		1113				
	650					
		1637				
		300				
				109127		
114600	2650	20911	—	181633	146	
		1637				
			6491			
114600	2650	22548	6491	181633	146	

Figure 3-2A

	Date	Receipts and Disbursements	Ck No	1 Income	2 Disbursements	3 Interest
1	—	(Brought Forward)	—	1257777	328068	—
2	3-9	L.C. Brown	502	142123		
3	3-10	17.9 Gals. Gasoline	—		2289	
4	3-12	Long Distance Telephone Call	827		136	
5	3-12	Printed Forms	829		6250	
6	3-17	City Tag for Vehicle	—		1000	
7	3-17	B.W. Baker	985	311550		
8	3-17	Office Supplies	842		9507	
9	3-17	16.1 Gals. Gasoline	—		2059	
10	3-17	March Invoices	843		142123	
11	3-19	Car Wash	—		300	
12	3-21	10.9 Gals. Gasoline	—		1394	
13	3-24	L.C. Brown	508	1041625		
14	3-24	Plumbing Material	844		162000	
15	3-25	Car Wash	—		300	
16	3-26	B.W. Baker	992	26854		
17	3-28	March Invoices	845		25858	
18	3-28	17.7 Gals. Gasoline	—		2264	
19	4-2	Typewriter Supplies	850	2779929	683548 / 6121	—
20	4-2	17.3 Gals. Gasoline	—		2213	
21	4-4	L.C. Brown	512	528357		
22	4-5	March Invoices	853		89732	
23	4-6	L.C. Brown	515	440000		
24	4-7	20.7 Gals. Gasoline	—		2648	
25	4-8	Car Wash	—		300	
26	4-9	Long Distance Telephone Calls	856		563	
27	4-12	Car Inspection	—		700	
28	4-14	17.3 Gals. Gasoline	—		2213	
29	4-15	L.C. Brown	524	440000		
30	4-16	8.2 Gals. Gasoline	—		1044	
	—	Total (Carry Forward)	—	4888286	789087	

Income and Disbursements

4	5	6	7	8	9	10
Taxes	Office	Vehicle	Insurance	Baker Job	Brown Job	
1146 00	26 50	225 48	64 91	181 633	146	
		22 89				
				136		
	62 50					
10 00						
	95 07					
		20 59			142 123	
		3 00				
		13 94				
					162 000	
		3 00				
				258 58		
1156 00	184 07	22 64	64 91	207 627	304 269	
	61 21	311 54				
		22 13				
				897 32		
		26 48				
		3 00				
				53		
		7 00				
		22 13				
		10 49				
1156 00	245 28	402 77	64 91	297 922	304 269	

Figure 3-2B

	DATE	RECEIPTS AND DISBURSEMENTS	Ck No.	1 INCOME	2 DISBURSEMENTS	3 INTEREST
1	—	(BROUGHT FORWARD)	—	4188286	789087	—
2	4-20	14.3 GALS. GASOLINE	—		1830	
3	4-25	F.I.C.A. (1st) Quarter	867		84789	
4	4-25	STATE F.U.T.A. (1st) Quarter	868		858	
5	4-25	BUSINESS AND OCCUPATION TAX	870		59174	
6	4-25	17.3 GALS. GASOLINE	—		2213	
7	4-27	E.J. GREEN	778	1023303		
8	4-28	INVOICE FOR CONCRETE	873		80798	
9	4-28	INVOICE FOR FILL DIRT	874		275130	
10	4-30	CAR WASH	—		300	
11	4-30	20.4 GALS. GASOLINE	—		2610	
12	5-2	OFFICE SUPPLIES	—	5211589	1296789 / 682	—
13	5-7	SERVICE AGREEMENT ON EQUIPMENT	883		27600	
14	5-7	20.5 GALS. GASOLINE	—		2622	
15	5-11	L.C. BROWN	539	996900		
16	5-11	E.J. GREEN	790	218645		
17	5-11	APRIL INVOICES	889		260857	
18	5-12	300 WATT BULBS (CONCRETE WORK)	—		568	
19	5-13	B.W. BAKER	1026	680092		
20	5-13	APRIL INVOICES	894		131523	
21	5-13	CONCRETE INVOICE	895		44298	
22	5-13	20.4 GALS. GASOLINE	—		2610	
23	5-14	SUPPLIES FOR TYPEWRITER	897		4784	
24	5-14	LONG DISTANCE TELEPHONE CALL	898		363	
25	5-20	PLUMBING SUPPLIES	902		51798	
26	5-20	18.7 GALS. GASOLINE	—		2392	
27	5-21	OFFICE SUPPLIES	—		879	
28	5-26	19.5 GALS. GASOLINE	—		2495	
29	5-28	ADDITIONAL INSURANCE PREMIUM	907		2700	
30	5-28	CONCRETE INVOICES	909		140333	
		Total (CARRY FORWARD)		7107226	1973293	—

Income and Disbursements

4	5	6	7	8	9	10
TAXES	OFFICE	VEHICLE	INSURANCE	BAKER Job	BROWN Job	GREEN Job
1156 00	245 28	402 77	64 91	2979 22	3042 69	—
		18 30				
847 89						
8 58						
591 74						
		22 13				
						807 98
						275 30
		3 00				
		26 10				
2604 21	245 28	472 30	64 91	2979 22	3042 69	3559 28
	6 82					
	276 00					
		26 22				
					1672 12	936 45
					5 68	
				1315 23		
				442 98		
		26 10				
	47 84					
				3 63		
					517 98	
		23 92				
	8 79					
		24 95				
			27 00			
					1403 33	
2604 21	584 73	573 49	91 91	4741 06	6641 80	4495 73

Figure 3-2C

	Date	Receipts and Disbursements	Ck No.	1 Income	2 Disbursements	3 Interest
1	—	(BROUGHT FORWARD)	—	7107226	1973293	
2	5-28	CAR WASH	—		300	
3	5-31	OFFICE SUPPLIES	913		7640	
4	6-1	MATERIAL FOR BAKER JOB		7107226	1981233 / 171	—
5	6-1	STEEL BEAMS	914		103896	
6	6-2	20.4 GALS. GASOLINE	—		2610	
7	6-8	20.9 GALS. GASOLINE	—		2673	
8	6-9	L.C. BROWN	557	885779		
9	6-10	P.O. BOX RENT	—		1500	
10	6-10	CAR WASH	—		300	
11	6-11	LONG DISTANCE CALLS	924		444	
12	6-11	CONTRACTORS LICENSE	925		5000	
13	6-13	15.3 GALS. GASOLINE	—		1957	
14	6-17	E.J. GREEN	844	1541937		
15	6-18	8.5 GALS. GASOLINE	—		1087	
16	6-20	JUNE INVOICES	934		823354	
17	6-22	B. W BAKER	1079	454909		
18	6-23	FILL DIRT	937		439750	
19	6-23	JUNE INVOICE	938		17435	
20	6-23	CONCRETE INVOICES	939		35438	
21	6-23	15.3 GALS. GASOLINE	—		1957	
22	6-28	E.J. GREEN	868	6116896		
23	6-28	HOUSE PACKAGE (Factory Built Home)	946		4688443	
24	7-1	17.5 GALS. GASOLINE	—	16106747	8107248 / 2238	—
25	7-2	CAR WASH	—		300	
26	7-2	INVOICES FOR STONE	947		27354	
27	7-9	LONG DISTANCE TELEPHONE CALLS	951		224	
28	7-9	JUNE INVOICE	956		4717	
29	7-9	CONCRETE INVOICE	957		35438	
30	7-9	17.0 GALS. GASOLINE	—		2175	
	—	TOTAL (CARRY FORWARD)		16106747	8179694	

Income and Disbursements

	4	5	6	7	8	9	10
	Taxes	Office	Vehicle	Insurance	Baker Job	Brown Job	Green Job
	260421	58473	57349	9191	474106	664180	449573
			300				
		7640					
	260421	66113	57649	9191	474106	664180	449573
					171		
							103896
			2610				
			2673				
		1500					
			300				
						265	179
	5000						
			1957				
			1087				
						147867	675487
							439750
					17435		
					35438		
			1957				
							4688443
	265421	67613	68233	9191	527150	812312	6357328
			2238				
			300				
						13026	14328
							224
							4717
						35438	
			2175				
	265421	67613	729466	9191	527150	860776	6376597

Figure 3-2D

	DATE	RECEIPTS AND DISBURSEMENTS	Ck. No.	1 INCOME	2 Disbursements	3 Interest
1		(Brought Forward)	—	16106747	8179694	—
2	7-11	Postage Stamps	—		2000	
3	7-16	17.0 Gals. Gasoline	—		2190	
4	7-16	Car Wash	—		300	
5	7-20	F.I.C.A. (2nd Quarter)	963		276141	
6	7-20	F.U.T.A. (2nd Quarter)	964		1721	
7	7-20	Business and Occupation Tax (2nd Qtr)	966		42078	
8	7-22	20.4 Gals. Gasoline	—		2630	
9	7-25	L.C. Brown	586	1042570		
10	7-25	June Invoices (Brown Job)	974		42954	
11	7-26	E.J. Green	907	2262804		
12	7-27	17.0 Gals. Gasoline	—		2190	
13	8-2	17.6 Gals. Gasoline	—	19412121	8551898 / 2270	—
14	8-2	Typewriter Supplies	985		4784	
15	8-5	Car Wash	—		300	
16	8-10	Tune-up on Vehicle	988		9500	
17	8-12	22.1 Gals. Gasoline	—		2850	
18	8-13	Office Supplies	994		2498	
19	8-16	8.1 Gals. Gasoline	—		1044	
20	8-20	22.6 Gals. Gasoline	—		2913	
21	8-20	Car Wash	—		300	
22	8-26	18.7 Gals. Gasoline	—		2410	
23	9-1	Liability Insurance	1011	19412121	8580767 / 211945	—
24	9-1	Insurance on Vehicle	1012		15385	
25	9-1	20.0 Gals. Gasoline	—		2578	
26	9-2	E.J. Green	962	2938040		
27	9-2	Invoice for Green Job	1013		19235	
28	9-8	18.7 Gals. Gasoline	—		2410	
29	9-10	Long Distance Telephone Call	—		367	
30	9-14	Invoices on Green Job	1023		45464	
		TOTAL (CARRY FORWARD)	—	22350161	8878151	—

Income and Disbursements

4 Taxes	5 Office	6 Vehicle	7 Insurance	8 Baker Job	9 Brown Job	10 Green Job
2654.21	676.13	729.46	91.91	527.50	8607.76	63765.97
	20.00					
		21.90				
		3.00				
276.41						
17.21						
420.78						
		26.30				
					398.67	30.87
		21.90				
5853.61	696.13	802.56	91.91	527.50	9006.43	63796.84
		22.70				
	87.84					
		3.00				
		95.00				
		28.50				
	24.98					
		10.44				
		29.13				
		3.00				
		24.10				
5853.61	768.95	1018.43	91.91	527.50	9006.43	63796.84
			2119.45			
		153.85				
		25.78				
						192.35
		24.10				
						3.67
						454.64
5853.61	768.95	1222.16	2211.36	527.50	9006.43	64447.50

Figure 3-2E

				1	2	3
	DATE	RECEIPTS AND DISBURSEMENTS	Ck No.	INCOME	DISBURSEMENTS	INTEREST
1	—	(BROUGHT FORWARD)	—	2235016 1	887815 1	—
2	9-15	Refund on Worker's Compensation	—		(8569)	
3	9-16	19.5 Gals. Gasoline	—		2515	
4	9-17	Car Licence (State)	1026		2000	
5	9-21	19.5 Gals. Gasoline	—		2515	
6	9-24	Service on Vehicle	1032		3365	
7	9-29	20.4 Gals. Gasoline	—		2630	
8	10-1	Car Wash	—	2235016 1	888260 7 / 300	—
9	10-3	Supplies for Copy Machine	1043		1 33 23	
10	10-4	E.J. Green	1010	23464 19		
11	10-4	September Invoices	1044		1406 01	
12	10-4	17.0 Gals. Gasoline	—		2190	
13	10-5	Inspection Sticker for Vehicle	—		700	
14	10-10	September Invoices	1050		5 32576	
15	10-10	20.4 Gals. Gasoline	—		2630	
16	10-10	F.I.C.A. Taxes (3rd Quarter)	1051		3 32287	
17	10-10	F.U.T.A. Taxes (3rd Quarter)	1052		231	
18	10-11	Long Distance Telephone Calls	1057		1 455	
19	10-11	Workers Compensation	1058		5 71914	
20	10-13	Supplies for Office Equipment	1060		7582	
21	10-15	Car Wash	—		300	
22	10-26	21.2 Gals. Gasoline	—		2710	
23	10-31	Concrete Invoice	1076		1 18260	
24	11-2	19.7 Gals. Gasoline	—	24696580	10609866 / 2520	—
25	11-8	E.J. Green	1071	29 42147		
26	11-9	October Invoices	1086		5 11481	
27	11-9	18.5 Gals. Gasoline	—		2365	
28	11-12	Long Distance Telephone Call	1091		302	
29	11-16	15.3 Gals. Gasoline	—		1951	
30	11-17	November Invoices	1097		67420	
	—	TOTAL (CARRY FORWARD)	—	2763827 7	11195911	—

Income and Disbursements

4	5	6	7	8	9	10
TAXES	OFFICE	Vehicle	Insurance	BAKER Job	BROWN Job	GREEN Job
585361	76895	122216	221136	527150	900643	6444750
			(8569)			
		2515				
2000						
		2515				
		3365				
		2630				
587361	76895	133241	212567	527150	900643	6444750
		300				
	13323					
						140601
		2190				
		700				
						532576
		2630				
332287						
	431					
						1455
			571914			
	7582					
		300				
		2710				
						118260
920079	97800	142071	784481	527150	900643	7237642
		2520				
						511481
		2365				
						302
		1957				
						67420
920079	97800	148913	794481	527150	900643	7816845

Figure 3-2F

Odometer Reading 12-31-XX: 40,518.9
(Total Mileage: 40,518.9 Less 32,363.4 = 8,155.5)

	Date	Receipts and Disbursements	Ck No.	1 Income	2 Disbursements	3 Interest
1	—	(Brought Forward)	—	2763872 7	11195911	
2	11-19	Service on Vehicle	1100		6715	
3	11-25	18.7 Gals. Gasoline	—		2392	
4	11-26	Service on Copy Machine	1104		6270	
5	11-29	E.J. Green	2105	871420		
6	12-2	20.0 Gals. Gasoline	—	28510147	11212 1288 / 2558	
7	12-7	State Contractors License	1116		5000	
8	12-8	17.1 Gals. Gasoline	—		2187	
9	12-10	Long Distance Telephone Calls	1124		861	
10	12-12	P.O. Box Rent	—		1500	
11	12-13	Car Wash	—		300	
12	12-13	15.6 Gals. Gasoline	—		1995	
13	12-15	E.J. Green	2126	612479		
14	12-16	December Invoices	1126		9048	
15	12-16	December Invoices	1127		12831	
16	12-16	Office Supplies	—		313	
17	12-17	17.3 Gals. Gasoline	—		2213	
18	12-23	15.3 Gals. Gasoline	—		1957	
19	12-23	Car Wash	—		300	
20	12-30	19.0 Gals. Gasoline	—		2430	
21	12-31	December Invoices	1139		82169	
22						
23						
24						
25						
26						
27						
28						
29						
30						
	—	Total For Year	—	29122626	11418050	—

Income and Disbursements

4	5	6	7	8	9	10
TAXES	OFFICE	Vehicle	INSURANCE	BAKER Job	BROWN Job	GREEN Job
920079	97800	148913	784481	527150	900643	7816845
		6715				
		2392				
	6270					
920079	104070	158020	784481	527150	900643	7816845
		2558				
5000						
		2187				
						861
	1500					
		300				
		1995				
						90148
						12831
	313					
		2213				
		1957				
		300				
		2430				
						82169
925079	105883	171960	784481	527150	900643	8002854

Figure 3-2G

	Date	Name	Ck No	1 Disbursements	2 Baker Job	3 Brown Job	4 Green Job
1	1-7	Walton Construction Co	761	297 00	297 00		
2	2-28	D+A Plumbing Co.	816	4835 00	4835 00		
3	3-24	Holston Drywall, Inc	840	1065 00		1065 00	
4	5-13	W.C. Martin, Masonry Contractor	893	534 33	534 33		
5	5-13	Walton Construction Co.	896	414 72		414 72	
6	6-2	W.C. Martin, Masonry Contractor	916	2780 00			2780 00
7	6-10	Holston Drywall, Inc	922	360 60 / 5936 65	1316 33	360 60 / 1840 32	2780 00
8	8-12	B+D Roofing Company	991	544 30			544 30
9	9-23	W.C. Martin, Masonry Contractor	1030	919 69			919 69
10	9-30	W.C. Martin, Masonry Contractor	1036	683 85 / 8084 49	1316 33	1840 32	683 85 / 4927 84
11	10-7	WC. Martin, Masonry Contractor	1047	675 00			675 00
12	10-14	W.C. Martin, Masonry Contractor	1063	267 30			267 30
13	10-24	W.C. Martin, Masonry Contractor	1067	677 70			677 70
14	10-28	W.C. Martin, Masonry Contractor	1070	215 20			215 20
15	10-31	Walton Construction Co.	1077	289 14 / 10208 83	1316 33	1840 32	289 14 / 7052 18
16	11-4	W.C. Martin, Masonry Contractor	1083	189 00			189 00
17	11-9	Holston Drywall, Inc	1087	2041 20			2041 20
18	12-9	W.C. Martin Masonry Contractor	1119	45 00			45 00
19							
20							
21							
22							
23							
24							
25							
26							
27							
28							
29							
30							
		Total For Year	—	12484 03	1316 33	1840 32	9327 38

Subcontracted work
Figure 3-3

SCHEDULE C
(Form 1040)

Department of the Treasury
Internal Revenue Service (0)

Profit or (Loss) From Business or Profession
(Sole Proprietorship)
Partnerships, Joint Ventures, etc., Must File Form 1065.
▶ **Attach to Form 1040 or Form 1041.** ▶ **See Instructions for Schedule C (Form 1040).**

OMB No. 1545-0074

19 X X
09

Name of proprietor	*Employer's Name*

Social security number of proprietor: *000 00 0000*

A Main business activity (see Instructions) ▶ *Services* ; product ▶ *Building*

B Business name and address ▶ ....

C Employer identification number: *5 4 0 5 6 8 7 5 5*

D Method(s) used to value closing inventory:
(1) ☐ Cost (2) ☐ Lower of cost or market (3) ☐ Other (attach explanation)

E Accounting method: (1) ☒ Cash (2) ☐ Accrual (3) ☐ Other (specify) ▶ ....

	Yes	No

F Was there any major change in determining quantities, costs, or valuations between opening and closing inventory? . . . .
If "Yes," attach explanation.

G Did you deduct expenses for an office in your home? . . . .

PART I.—Income

1 a Gross receipts or sales	1a	*291,226 26*
b Less: Returns and allowances	1b	
c Subtract line 1b from line 1a and enter the balance here	1c	
2 Cost of goods sold and/or operations (Part III, line 8) . .	2	
3 Subtract line 2 from line 1c and enter the **gross profit** here . .	3	
4 a Windfall Profit Tax Credit or Refund received in 1983 (see Instructions)	4a	
b Other income	4b	
5 Add lines 3, 4a, and 4b. This is the **gross income** . . . . ▶	5	*291,226 26*

PART II.—Deductions

6 Advertising			**23** Repairs		
7 Bad debts from sales or services (Cash method taxpayers, see Instructions) .			**24** Supplies (not included in Part III) .		*94,306 47*
8 Bank service charges			**25** Taxes (Do not include Windfall Profit Tax here. See line 29.)		*9,296 53*
9 Car and truck expenses . . .	*1,719 60*		**26** Travel and entertainment . . .		
10 Commissions			**27** Utilities and telephone		*168 27*
11 Depletion			**28 a** Wages	*130,555 00*	
12 Depreciation and Section 179 deduction from Form 4562 (not included in Part III) . . .			**b** Jobs credit		
			c Subtract line 28b from 28a . .		*130,555 00*
13 Dues and publications			**29** Windfall Profit Tax withheld in 1983		
14 Employee benefit programs . . .			**30** Other expenses (specify):		
15 Freight (not included in Part III) . .			**a** *Subcontracts*		*12,484 03*
16 Insurance	*7,877 26*		**b**		
17 Interest on business indebtedness . .	*118 16*		**c**		
18 Laundry and cleaning			**d**		
19 Legal and professional services . .			**e**		
20 Office expense	*1,058 83*		**f**		
21 Pension and profit-sharing plans . .			**g**		
22 Rent on business property			**h**		
			i		

31 Add amounts in columns for lines 6 through 30i. These are the **total deductions** . . . . . ▶	31	

32 Net profit or (loss). Subtract line 31 from line 5 and enter the result. If a profit, enter on Form 1040, line 12, and on Schedule SE, Part I, line 2 (or Form 1041, line 6). If a loss, go on to line 33 . . . . . . | 32 | |

33 If you have a loss, you must answer this question: "Do you have amounts for which you are not at risk in this business (see Instructions)?" ☐ Yes ☐ No
If "Yes," you must attach Form 6198. If "No," enter the loss on Form 1040, line 12, and on Schedule SE, Part I, line 2 (or Form 1041, line 6).

PART III.—Cost of Goods Sold and/or Operations (See Schedule C Instructions for Part III)

1 Inventory at beginning of year (if different from last year's closing inventory, attach explanation)	1	
2 Purchases less cost of items withdrawn for personal use	2	
3 Cost of labor (do not include salary paid to yourself)	3	
4 Materials and supplies	4	
5 Other costs	5	
6 Add lines 1 through 5	6	
7 Less: Inventory at end of year	7	
8 **Cost of goods sold and/or operations.** Subtract line 7 from line 6. Enter here and in Part I, line 2, above.	8	

For Paperwork Reduction Act Notice, see Form 1040 Instructions.

Schedule C (Form 1040)

Schedule C (Form 1040)
Figure 3-4

Preparing Schedule C (Form 1040)

When you prepare your annual income taxes, you'll be required to fill out Schedule C (Form 1040). You'll already have most of the information you need. Just copy it from the income and disbursements record (Figure 3-2G), the subcontracted work record (Figure 3-3), and the consolidated report of labor costs per job (Figure 2-4F in Chapter 2).

Figure 3-4 is a sample of Schedule C (Form 1040). This report includes the following information:

1) Gross receipts— This is your total annual income. Take this information from the last line of column 1 in Figure 3-2G. In our example, gross receipts come to $291,226.26. Enter this amount on line 1.

2) Car and truck expenses— For your total annual vehicle expenses, look at column 6 in Figure 3-2G. This comes to $1,719.60. Enter this amount on line 9.

3) Insurance— This is a combination of the total insurance shown in column 7 in Figure 3-2G ($7,844.81) plus your prorated insurance for maintaining an office in the home ($32.45). Total insurance comes to $7,877.26. Enter this amount on line 16.

4) Interest— If you maintain an office in the home, you can prorate the interest on your home mortgage. The prorated interest comes to $118.16. Enter this amount on line 17. Remember that you must also deduct this $118.16 from home mortgage interest on Schedule A (Form 1040).

5) Office expenses— For your total annual office expenses, look at column 5 in Figure 3-2G. The total comes to $1,058.83. Enter this amount on line 20.

6) Materials and supplies— To calculate your total annual materials and supplies, add the totals from columns 8, 9 and 10 in Figure 3-2G. The grand total comes to $94,306.47. Enter this amount on line 24.

7) Taxes— These are a combination of the taxes shown in column 4 in Figure 3-2G ($9,250.79) plus prorated real estate taxes ($45.74). The combined total comes to $9,296.53. Enter this amount on line 25.

The prorated real estate taxes are proportional to the percentage of space occupied by the office in your home. For example, if the office occupies 11% of the total house space, then 11% of your real estate taxes can be charged to business expense. In our example, real estate taxes came to $415.86. The office occupied 11% of the total house space. This gives us prorated real estate taxes of $45.74.

Remember that you must also deduct this $45.74 from your real estate taxes on Schedule A (Form 1040).

8) Utilities and telephone— If you maintain an office in your home, you can deduct a percentage of the total utility costs. In our example, this comes to $168.27. Enter this amount on line 27.

9) Wages— The total annual cost of labor comes from your consolidated report of labor costs per job. (See Figure 2-4F in Chapter 2.) In our example, the total cost of labor comes to $130,555.00. Enter this amount on line 28A.

10) Subcontracted work— For your total annual cost of subcontracted work, look at column 1 in Figure 3-3. In our example, the total comes to $12,484.03. Enter this amount on line 30A in Figure 3-4.

Now you know how to keep accurate cost records for payroll, billing, income and disbursements, and subcontracted work. And we've seen how these records can help with your quarterly reports, year-end reports and income taxes. In the next chapter, we'll move directly to the building site and set up some records that will save you money on site preparation.

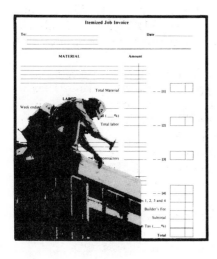

Chapter 4

Site Preparation Records

Up to this point, we've covered record-keeping for payroll, billing, income and disbursements, and subcontracted work. Now we'll explain how to keep cost records for each phase of the construction project: site preparation, concrete, finishes, etc. Each type of work is different and requires its own record-keeping procedure.

Remember the prime reason for keeping accurate cost records: Your actual costs of completed work are the most reliable guide for estimating future jobs. Cost records in your files show *your* material and manhour costs for *your* kind of work in *your* area and with *your* crews. No estimating manual or "standard" cost can do this.

In this chapter, we'll look at the costs associated with preparing the site for construction. These include: preliminary work, site clearing, excavation, fill dirt, site cleaning and hauling.

Preliminary Work

Every job has certain preliminary costs that should be included in your cost estimate. These include: the architect's fee, plot plans, building permit, water connection, sewer connection, temporary phone and electric service. These costs will be different for each job. So your cost records won't be of much help here except as a checklist of items that must be included. Once you've overlooked an item that costs several hundred or thousand dollars, you'll have no trouble remembering that item on future estimates.

Figure 4-1 shows how to record your estimated and actual costs for preliminary work. It's fairly easy to estimate these costs. Let's look at each of them in detail.

Architect's Fee
The architect's fee (or the cost of stock plans) may or may not be part of your construction costs. If you're a spec builder who builds houses for sale, they will be. If you're building a house on a contract or cost-plus basis, the plans will be supplied by the owner. If the owner pays for the blueprints, record this as shown in Figure 4-1.

Plot Plans
Plot plans are essential on every project. They're especially helpful when estimating construction costs. Normally, the owner buys the plot plans. If you prepare the plot plans yourself, include this expense as part of your construction costs. If the owner pays for the plot plans, record this as shown in Figure 4-1.

Building Permit
Nearly anywhere you build, alter or enlarge a structure, you'll need authorization from some governmental authority. The building code and zoning ordinances regulate construction details, the size and the type of building that can be built in most areas. Any time the government regulates building activity, there will be a fee schedule to show you the cost of regulation.

Site Preparation

Date: _7-2-XX_   _BROWN_ job

	Estimated Costs	Actual Costs
1) Preliminary work:		
Architect's fee or stock plans	$ *By Owner*	$ *—0—*
Plot plans	$ *By Owner*	$ *—0—*
Building permit	$ *108.00*	$ *108.00*
Water connection	$ *250.00*	$ *250.00*
Temporary water service	$ *35.00*	$ *32.50*
Sewer connection	$ *250.00*	$ *250.00*
Temporary phone and electric service	$ *65.00*	$ *69.78*
2) Site clearing	$ *N/A*	$ *—0—*
3) Excavation	$ *250.00*	$ *262.50*

Note: Actual cost $ _262.50_ divided by _155_ cu. yds. = $ _1.69_ per cu. yd.

Blasting required: (Yes) - ((No))

Weather conditions: ((Good)) - (Fair) - (Poor)

4) Fill dirt (*Includes spreading and compaction*)	$ *3,000.00*	$ *3,000.00*

Note: Actual cost $ *3,000.00* divided by *1,685* cu. yd. = $ *1.78* per cu. yd.

5) Site cleaning and hauling	$ *100.00*	$ *95.00*

Site Preparation
Figure 4-1

The cost of building permits and inspection fees is usually based on the estimated value of the structure. Many cities and counties use "standard" costs per square foot of building. The permit-issuing authority will explain the fee schedule over the phone. Remember to record this important cost. On most residential jobs, the building permit and inspection fee will not be more than 1% of the construction cost.

In Figure 4-1, the builder paid for the building permit. The estimated cost was $108.00. The actual cost was also $108.00.

Water Connection and Temporary Water Service
If you plan to rely on a public agency for water at the site, there's sure to be a connection charge or deposit required before the first drop of water flows. The cost may be as little as a hundred dollars if water is already available at the site. But it could be several hundred or even thousands if the main does not run by the property.

The cost of water actually used during construction will probably be insignificant. It will be less than the connection charge, unless you need a lot of water for compacting soil. But many water districts have minimum monthly charges. This minimum charge becomes significant if the job runs more than a few months.

If there's no public water available, you'll have to include the cost of a well or piping to a spring. Be sure you know who is locally responsible for maintaining this private water supply and what the total costs will be. Operating the private water supply will be your responsibility during construction.

Sometimes you'll have a job where no water is available during construction. In this case, you'll have to truck water to the site and make it available for your crews and subcontractors. Most of your subcontractors will assume that water and electricity are supplied by the general contractor. They'll set their bids accordingly.

Avoid unpleasant surprises. Make sure water is going to be available, and determine the cost. Be sure to allow for extra water used by brick masons and drywall installers. In Figure 4-1, the estimated cost and the actual cost for water connection were both $250.00. The estimated cost for temporary water service was $35.00. The actual cost was $32.50.

Sewer Connection
If sewage disposal is to a public sewer line, you'll have the expense of a sewer tap. You'll have to pay a connection charge based on the number of bathrooms or type of occupancy and building size.

This fee may be thousands of dollars. The sewer district will estimate this cost and advise you if your connection might be delayed for any reason. No certificate of occupancy can be issued (in most communities) until the sewer line is connected. And the sewer line won't be connected until the fees are paid.

Disposal to a private sewage system, such as a septic tank, is possible in many areas. But it's prohibited in most urban communities. If you intend to rely on a septic system or a privately owned disposal system that serves two or more families, get firm bids that include the cost of maintaining the system during construction.

In Figure 4-1, the established fee for this connection was $250.00. So the estimated cost and the actual cost were the same.

Temporary Phone and Electric Service
Your phone and utility companies have fee schedules for installing temporary service. There will also be monthly service charges. These are part of your construction costs. On some jobs, you'll have to provide a drop pole and temporary electrical distribution panel until permanent service equipment is installed. You'll have to pay for installation, power used during construction, and pick-up when the job is finished. In cold weather, remember to allow extra for providing temporary heat for your workers.

In Figure 4-1, the estimated cost of temporary phone and electric was $65.00. The actual cost was $69.78.

This completes our look at preliminary work. Now let's go on to site clearing, excavation, fill dirt, site cleaning and hauling.

Site Clearing
Nearly all construction begins with removal of trees, brush, old buildings, pavement, fences, poles and debris. These must all be cleared away before work can start. When the amount of site clearing is minimal, you may want to handle the work yourself. But if site clearing is a major item, get a firm bid from an experienced subcontractor.

On small jobs, site clearing may be less than a full day's work for the crew and equipment you hire. Plan to have other productive work the crew can do in case they finish site-clearing before the work day is done. Otherwise, you'll end up paying out a full day's pay for only part of a day's work.

In Figure 4-1, no site clearing was required. This is indicated in both the estimated cost and actual cost columns.

Excavation

The construction of any foundation will require some excavation. The quantity of earth to be moved depends on the design of the house, the frost level, the soil type, and the size of the foundation, basement, crawl space, or slab.

Most residential builders subcontract excavation work, except for finish grading and light trenching. If you decide to do your own excavation work, here's what you'll need to know.

The cost of excavation depends primarily on soil conditions, the type of equipment and the weather. If rock has to be ripped or blasted, expect production to drop to a fraction of what it would otherwise be. If hand work is needed around obstructions, you'll only move about one cubic yard per manhour. Cold and wet weather can increase costs by 15% or more.

Larger, more expensive equipment generally lowers excavation costs. But it isn't economical to use large equipment on short jobs or smaller sites.

Be sure to identify in your estimate the type of equipment you plan to use and the type of soil you expect. When the job is finished, note the equipment actually used, the actual type of soil moved and any weather or other problems that changed productivity rates.

Figure 4-1 shows you how to compute the actual unit cost of excavation. For example, the estimated cost of the excavation was $250.00. The actual cost to excavate 155 cubic yards was $262.50. This includes transporting the earth-moving equipment to the job site. Divide the actual cost ($262.50) by the number of cubic yards (155), and this gives you the unit cost of $1.69 per cubic yard.

No blasting was required. Be sure to indicate this as shown on Figure 4-1. The weather was good. This is also indicated on Figure 4-1. When you prepare your next bid for excavation with the same equipment and for similar conditions, you can begin with a base price of $1.69 per cubic yard.

Figure 4-2A shows a weekly time sheet for excavation work. This work is described in the daily log at the bottom of the page.

Fill Dirt

The starting grade elevations, the finish grade elevations, and the location of the building on the site all determine the quantity of fill required. Remember that earth swells about 25% when ex-cavated. And it must be compacted to near its original density to prevent subsidence after work is completed.

Common earth and loam swell and compact about 25%. Sandy soil swells less. Common sand swells only about 15%. But moist clay swells about 30% and dry clay swells even more, about 35%.

Fill dirt may not be available on the job site. If the fill is imported, and compaction is going to be a major cost item, the contract with the fill supplier must provide the following information:

• Grade of dirt: no rock, 10% rock maximum, no shale, etc.

• Degree of compaction: 6-inch layers, 8-inch layers, continuous

• Elevation of finish grade after compaction is completed

• Cost of compaction per cubic yard

Figure 4-1 shows the estimated cost of fill dirt at $3,000.00. The plot plan told us we needed 1,685 cubic yards of soil. This included a 25% compaction allowance. Our $3,000.00 cost divided by 1,685 gives us a unit cost of $1.78 per cubic yard. Use this unit cost as a base price for estimating future fill dirt requirements.

Figures 4-2B through 4-2D show weekly time sheets covering the fill dirt work on the job. The work is described in the daily log at the bottom of each page.

Site Cleaning and Hauling

Finally, you'll need to estimate the cost of disposing of waste and rubble, cleaning the area, and hauling equipment, scaffolding and reuseable materials away from the job site. These are *significant* costs, and they should *not* come out of your profit. Keep accurate cost records on these items so that you have a sound basis for making future estimates.

Our estimated cost for site cleaning was $100.00, as shown in Figure 4-1. Our actual cost was $95.00.

In this chapter, we've learned how to keep accurate cost records for the first phase of the construction project — site preparation. In Chapter 5, we'll learn how to save money on the next phase of construction — footings.

Weekly Time Sheet

For period ending _8-21-XX_   _BROWN_ job

	Name	Exemptions	Days *AUGUST*						Rate	Hours worked		Total earnings
			16 M	17 T	18 W	19 T	20 F	21 S		Reg.	Over-time	
	Subcontractor											
1	C&S Construction Co.		✓	—	—	—	✓	—				
2												
3												
4												
5												
6												
7												
8												
9												
10												
11												
12												
13												
14												
15												
16												
17												
18												
19												
20												

Daily Log

Monday _Excavation for basement (155 cubic yards)_

Tuesday _—_

Wednesday _—_

Thursday _—_

Friday _Excavation and covered trench for water line (90 feet)_

Saturday _—_

Weekly Time Sheet
Figure 4-2A

Weekly Time Sheet

For period ending _4-9-XX_   _BROWN_ job

	Name	Exemptions	Days April						Rate	Hours worked		Total earnings
			4 M	5 T	6 W	7 T	8 F	9 S		Reg.	Over-time	
	Subcontractor											
1	C & S Construction Co		—	—	—	—	✓	✓				
2												
3												
4												
5												
6												
7												
8												
9												
10												
11												
12												
13												
14												
15												
16												
17												
18												
19												
20												

Daily Log

Monday _—_

Tuesday _—_

Wednesday _—_

Thursday _—_

Friday _Fill dirt and dozer work to level dirt and compact_

Saturday _Fill dirt_

Weekly Time Sheet
Figure 4-2B

Weekly Time Sheet

For period ending ___4-16-XX___   _Brown_ job

	Name Subcontractor	Exemptions	Days April						Rate	Hours worked		Total earnings
			11 M	12 T	13 W	14 T	15 F	16 S		Reg.	Over-time	
1	C+S Construction Co.		✓	✓	—	—	—	—				
2												
3												
4												
5												
6												
7												
8												
9												
10												
11												
12												
13												
14												
15												
16												
17												
18												
19												
20												

Daily Log

Monday _Fill dirt_

Tuesday _Dozer work to level dirt and compact_

Wednesday _—_

Thursday _—_

Friday _—_

Saturday _—_

Weekly Time Sheet
Figure 4-2C

Weekly Time Sheet

For period ending __4-30-XX__   __BROWN__ job

	Name	Exemptions	April 25 M	26 T	27 W	28 T	29 F	30 S	Rate	Reg.	Over-time	Total earnings
			Days.							**Hours worked**		
	Subcontractor											
1	C + S Construction Co		—	—	—	—	—	✓				
2												
3												
4												
5												
6												
7												
8												
9												
10												
11												
12												
13												
14												
15												
16												
17												
18												
19												
20												

Daily Log

Monday —

Tuesday —

Wednesday —

Thursday —

Friday —

Saturday *Fill dirt and dozer work to level dirt and compact*

Weekly Time Sheet
Figure 4-2D

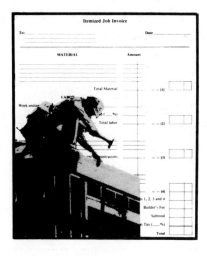

Chapter 5

Footing Records

It isn't always easy to estimate the cost of excavating, forming, pouring, and stripping the forms for a footing. There's more to consider than just the size of the foundation. You also have to think about soil conditions, the type of footing required, footing location, and weather conditions.

Calculating material requirements is no problem. There isn't much waste. But estimating footing labor is more difficult. The only accurate way is to use your own cost records as a guide.

In this chapter, we'll learn how to make accurate footing estimates. We'll show you the easy way to compute the materials and labor you'll need. We'll calculate the requirements for concrete, crushed stone, rebar, and the manhours required to hand-excavate, form, pour concrete and strip forms.

Footing Materials
The most commonly-used footing materials are concrete, crushed stone and rebar. Let's look at each of these materials in detail.

Concrete
A conventional rule of thumb for estimating concrete volume is:

Length of footing in feet, multiplied by the width in decimal equivalents of a foot, multiplied by the depth in decimal equivalents of a foot, divided by 27.

For example, if a footing is 234' long, 24" wide, and 9" deep, how many cubic yards of concrete are required?

$$\frac{234.0' \times 2.0' \,(24\text{" width}) \times 0.75' \,(9\text{" depth})}{27}$$

= 13.0 cu. yds.

Figure 5-1 shows you an easier way to estimate concrete volume. This chart will save you time and prevent expensive mistakes. Here's how it works:

1) Look at the 24" column under footing width.

2) Look down this column to the 9" footing depth.

3) Where these two lines intersect, find the factor 0.05556.

4) Multiply this factor (0.05556), by the total linear feet (234). This will give you the number of cubic yards of concrete required, (13).

The factors in Figure 5-1 are precise enough for any job, and they include most of the footing widths you're likely to pour.

Let's try another one: The footing for a small building is 108' long, 16" wide and 6" deep.

$$\frac{108.0' \times 1.333' \,(16\text{" width}) \times 0.5' \,(6\text{" depth})}{27}$$

= 2.67 cu. yds.

Footing depth	Footing width									
	12"	14"	16"	18"	20"	22"	24"	26"	28"	30"
6"	.01852	.02160	.02469	.02778	.03086	.03395	.03704	.04012	.04321	.04630
7"	.02160	.02521	.02881	.03241	.03601	.03961	.04321	.04681	.05041	.05401
8"	.02469	.02881	.03292	.03704	.04115	.04527	.04938	.05350	.05761	.06173
9"	.02778	.03241	.03704	.04167	.04630	.05093	.05556	.06019	.06481	.06944
10"	.03086	.03601	.04115	.04630	.05144	.05658	.06173	.06687	.07202	.07716
11"	.03395	.03961	.04527	.05093	.05658	.06224	.06790	.07356	.07922	.08488
12"	.03704	.04321	.04938	.05556	.06173	.06790	.07407	.08025	.08642	.09259
13"	.04012	.04681	.05350	.06019	.06687	.07356	.08025	.08693	.09362	.10031
14"	.04321	.05041	.05761	.06481	.07202	.07922	.08642	.09362	.10082	.10802
15"	.04630	.05401	.06173	.06944	.07716	.08488	.09259	.10031	.10802	.11574

Factors for concrete and crushed stone
Figure 5-1

The factor for this footing, taken from Figure 5-1, is 0.02469. Multiply the factor (0.02469), by the total linear feet (108). This gives you the number of cubic yards of concrete required, (2.67).

What if the footings are wider than the widths shown in Figure 5-1? Simply add factors together to get the correct width. For example, if a chimney footing is 6' long, 36" wide and 12" deep, how many cubic yards of concrete are required? Since there is no factor for a 36" width, let's combine the factors for the 20" width and the 16" width.

1) The factor for 20" x 12" is 0.06173

2) The factor for 16" x 12" is 0.04938

3) The factor for 36" x 12" is 0.11111

(0.06173 + 0.04938 = 0.11111)

Multiply the factor (0.11111), by the total linear feet (6). This gives you the number of cubic yards of concrete required, (0.67).

Crushed Stone
When crushed stone is required for a footing, Figure 5-1 will help you compute the number of cubic yards you'll need.

Let's say that you need 4" of crushed stone under a footing that is 124' long, 24" wide and 8" deep. To find the factor for 4" of crushed stone, take the factor for 24" x 8" and multiply it by 0.5.

1) The factor for 24" x 8" is 0.04938

2) 0.04938 (factor) x 0.5 = 0.02469 (factor for 4" crushed stone)

Then multiply the new factor (0.02469), by the total linear feet (124). This gives you the number of cubic yards of crushed stone required, (3.06).

To compute the tons of crushed stone required, multiply the number of cubic yards (3.06) by 1.35 (special factor for crushed stone). This will give you the number of tons of crushed stone required, (4.13).

Rebar
Figure 5-2 is a chart for computing rebar requirements. Like Figure 5-1, it will save you time and prevent errors in your estimates. The chart shows bar diameter, bar number, and bar weight in pounds per foot. If the rebar is sold by weight, here's a rule of thumb you can use to compute the total weight:

Multiply the total linear feet times the weight factor shown in Figure 5-2.

For example, what is the total weight of 1,040 linear feet of number 4 rebar?

1,040 (linear feet) x 0.668 (weight factor) = 694.72 lbs.

Note: round this off to 695 lbs.

Bar diameter (inches)	Bar number	Bar weight (pounds per foot)
¼''	2	.167
⅜''	3	.376
½''	4	.668
⅝''	5	1.043
¾''	6	1.502
⅞''	7	2.044
1''	8	2.670

Factors for rebar
Figure 5-2

In addition to concrete, crushed stone and rebar, be sure to include in your estimate: form lumber, stakes, and additives to protect the concrete in cold weather.

We've seen the easy way to compute material requirements for concrete, crushed stone and rebar. Now let's look at how to make accurate footing labor estimates.

Footing Labor

Your weekly time sheets and daily logs are the key to accurate footing labor estimates. They'll show you the manhours required per cubic yard of footing work. Let's look at some examples.

Figures 5-3A and 5-3B are weekly time sheets for the Baker job. This job was a large two-story house, with a crawl space, an attached two-car garage, and no basement.

The daily log indicates that work began on Monday, June 14. The footing work was completed on Tuesday, June 22. It took a total of 123 manhours to hand-excavate and pour 25 cubic yards of concrete. Divide the total manhours (123) by the cubic yards of concrete required (25). This gives you the number of manhours required per cubic yard of footing work, (4.92).

Now look at Figures 5-4A and 5-4B. These are weekly time sheets for the Brown job. This job was a two-story house, but the foundation was smaller than the foundation for the Baker house. The Brown house had a basement and an attached two-car garage. Forms were used for the footing in the basement area. Due to a sloping subgrade, a stepped footing was required to distribute the load of the wall.

The daily log indicates that footing work began on Wednesday, August 18. Work ended on Friday, August 27. A total of 154 manhours were required to lay out the footing, hand-excavate, form the basement area and the stepped footing, pour 14

cubic yards of concrete, and strip the forms from the footing. This job required 11 manhours per cubic yard of footing work.

Figure 5-5A is a footing labor worksheet that will help you compute the manhours required to hand-excavate, pour concrete and strip forms. Take the total manhours from the weekly time sheets. For example, the footing work manhours for the Baker job totaled 123. Record this number on your worksheet, as shown in Figure 5-5A. Then record the total cubic yards of concrete used on the job. For the Baker job, it was 25 cubic yards. The total manhours (123) divided by cubic yards of concrete (25) gives you the number of manhours required per cubic yard of concrete (4.92). This number is called the manhour factor.

To estimate *manhours* for footings on future bids, multiply the manhour factor (from similar jobs) times the estimated number of cubic yards of concrete. For example, if your manhour factor is 4.92 and the estimated number of cubic yards of concrete is 18, then the total manhours required to do the footing work will be 88.56. Round this off to 89 manhours.

To estimate *labor costs* for footings on future bids, multiply the total manhours times the labor rate per hour. In our example, the total manhours (89) multiplied by the labor rate per hour ($25.00) gives us the total labor cost for footing work ($2,225.00).

Remember to make allowances for abnormal working conditions. Winter weather can increase footing labor costs 10% to 15%. Rock removal will add to your costs. So will stepped footings or special forms. If concrete can't be unloaded directly from the truck into the footing, this will also raise your costs. Be sure to allow for these costs in your estimate.

Let's look at another example. Figure 5-5B is the manhour worksheet for the Green job. Let's say you need an estimated 20 cubic yards of concrete for a job you're bidding that's similar to the Green job. Multiply the manhour factor from the Green job (9.42) times the estimated cubic yards of concrete (20). This will give you a safe manhour estimate (188.4) for the new job. Round this off to 189 manhours.

To estimate the labor cost for the new job, multiply the total manhours (189) times the labor rate per hour ($25.00). This will give you the total labor cost for the footing work ($4,725.00).

Keeping good cost records will save you time and money. This means extra profit in your pocket. In this chapter, we've seen how to use your cost records to make accurate footing estimates. In the next chapter, we'll look at record-keeping for foundations.

Weekly Time Sheet

Page____of____pages

For period ending 6-19-XX

BAKER job

	Name	Exemptions	Days JUNE 14 M	15 T	16 W	17 T	18 F	19 S	Rate	Hours worked Reg.	Over-time	Total earnings
1	D. L. WEST		3½	8	5½	8	8	X		33		
2	J. E. KING		3½	8	5½	8	8	X		33		
3	A. L. KING		X	8	5½	8	8	X		29½		
4												
5												
6												
7												
8												
9												
10												
11												
12												
13												
14												
15												
16												
17												
18												
19												
20												

Daily Log

Monday *LAID OUT FOOTING FOR FOUNDATION*

Tuesday *WORK ON FOOTING*

Wednesday *FOOTING --- RAIN*

Thursday *FOOTING*

Friday *FOOTING FINISHED --- PASSED BY INSPECTOR --- READY FOR CONCRETE*

Saturday *XXXX*

Weekly time sheet - Baker job
Figure 5-3A

Weekly Time Sheet

Page____of____pages

For period ending 6-26-XX

Baker job

	Name	Exemptions	Days JUNE 21 M	22 T	23 W	24 T	25 F	26 S	Rate	Hours worked Reg.	Over-time	Total earnings
1	D.L. WEST		X	9½	7½	4	2½	X		23½		
2	A.L. KING		X	9	X	X	X	X		9		
3	J.E. KING		X	9	7½	X	X	X		16½		
4												
5												
6												
7												
8												
9												
10												
11												
12												
13												
14												
15												
16												
17												
18												
19												
20												

Daily Log

Monday ____ RAIN

Tuesday ____ POURED FOOTING (25 cu. yds.)---WEATHER CONDITIONS: Muddy

Wednesday ____ LAID OUT HOUSE CORNERS FOR BRICK MASONS

Thursday ____ TEMPORARY GRAVEL ON DRIVEWAY

Friday ____ Blocks, MORTAR AND SAND DELIVERED to Job SITE

Saturday ____ XXXX

Weekly time sheet - Baker job
Figure 5-3B

Weekly Time Sheet

Page____of____pages

For period ending _8-21-XX_   _BROWN_ job

	Name	Exemptions	Days August 16 M	17 T	18 W	19 T	20 F	21 S	Rate	Reg.	Over-time	Total earnings
1	D.L. White		X	X	3	8	8	X		19		
2	J.E. King		X	X	3	8	8	X		19		
3	N.R. Farlow		X	X	3	8	8	X		19		
4												
5												
6												
7												
8												
9												
10												
11												
12												
13												
14												
15												
16												
17												
18												
19												
20												

Daily Log

Monday XXXXX

Tuesday XXXXX

Wednesday LAID OUT FOOTING FOR FOUNDATION

Thursday FORMS FOR FOOTING IN BASEMENT AREA

Friday Footing

Saturday XXXXX

Weekly time sheet - Brown job
Figure 5-4A

Weekly Time Sheet

Page____of____pages

For period ending __8-28-XX__   __BROWN__ job

	Name	Exemptions	Days AUGUST						Rate	Hours worked		Total earnings
			23 M	24 T	25 W	26 T	27 F	28 S		Reg.	Over-time	
1	D. L. White		8	8	6	6½	6½	X		31		
2	J. E. King		8	8	6	6½	2½	X		31		
3	W. R. Farlow		8	8	6	6½	X	X		28½		
4	F. N. Neal		X	X	X	6½	X	X		6½		
5												
6												
7												
8												
9												
10												
11												
12												
13												
14												
15												
16												
17												
18												
19												
20												

Daily Log

Monday __Footing --- Stepped footing in garage area__

Tuesday __Footing --- Stepped footing in garage area__

Wednesday __Finished footing --- Passed by inspector --- Ready for concrete__

Thursday __Poured concrete for footing (14 cu. yds.) Wheelbarrows required__

Friday __Removed forms from footing__

Saturday __XXXXX__

Weekly time sheet - Brown job
Figure 5-4B

Date: **6-22-XX**   **BAKER** job

Manhour factor:

Total manhours divided by cubic yards concrete = Manhour factor

Total manhours **123** ÷ cubic yards concrete **25** = Manhour factor **4.92**

Weather conditions: Good (Fair) Poor

Rock removal required: Yes (No)

Wheelbarrows required: Yes (No)

Estimating manhours for future bids:

Manhour factor multiplied by estimated cubic yards concrete = Total manhours

Manhour factor **4.92** x estimated cubic yards concrete **18** = Total manhours **88.56**

Estimating labor costs for future bids:

Total manhours multiplied by rate per hour = Estimated labor cost

Total manhours **89** x rate per hour **$25.⁰⁰** = Estimated labor cost $**2,225.⁰⁰**

Footing Labor Worksheet - Baker Job
Figure 5-5A

Date: 4-26-XX

GREEN job

Manhour factor:

Total manhours divided by cubic yards concrete = Manhour factor

Total manhours ___179___ ÷ cubic yards concrete ___19___ = Manhour factor ___9.42___

Weather conditions: (Good) - Fair - Poor

Rock removal required: Yes - (No)

Wheelbarrows required: Yes - (No)

Estimating manhours for future bids:

Manhour factor multiplied by estimated cubic yards concrete = Total manhours

Manhour factor ___9.42___ x estimated cubic yards concrete ___20___ = Total manhours ___188.4___

Estimating labor costs for future bids:

Total manhours multiplied by rate per hour = Estimated labor cost

Total manhours ___188___ x rate per hour $___25.00___ = Estimated labor cost $___4725.00___

Footing Labor Worksheet - Green Job
Figure 5-5B

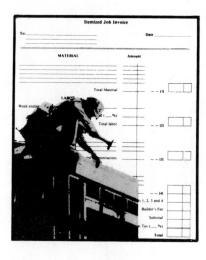

Chapter 6

Foundation Records

Estimating is faster and more accurate when you can reduce your mathematical calculations to a minimum. Factor tables help you do this. In this chapter, we'll learn how to make quick, accurate estimates for foundation work. With our cost records as our guide, we'll use factor tables to help us calculate both materials and labor.

Foundation Materials

Building a foundation requires: crushed stone, concrete block, masonry cement and sand, and parging materials (mortar, bituminous coating and drain tile). Let's take a look at each of these materials.

Crushed Stone

If you have a basement area 65'6'' x 32'6'' x 4'' deep, how many cubic yards of crushed stone will you need? And how many tons of crushed stone should you order?

The mathematical formula for computing cubic yard requirements is:

Length in feet and decimal equivalents of a foot, multiplied by the width in feet and decimal equivalents of a foot, multiplied by the depth (thickness) in decimal equivalents of a foot, divided by 27.

In our example:

$$\frac{65.5' \ (65'6'') \times 32.5' \ (32'6'') \times 0.33333' \ (4'')}{27}$$
$$= 26.3 \text{ cu. yds.}$$

The formula for converting cubic yards to tons (assuming one cubic yard weighs 2,700 lbs.) is:

Multiply the number of cubic yards times 2,700. Then divide by 2,000.

$$\frac{26.3 \text{ cu. yds. x 2,700 lbs.}}{2,000} = 35.5 \text{ tons}$$

Now let's do it the easy way. Here's how to make the same computation using the factor table shown in Figure 6-1. Look down the first column until you see the correct thickness (4''). Read across this line from left to right. The next column gives you the factor (0.01235). Multiply this factor times the area of the basement (2,128.75 SF). This will give you the number of cubic yards of crushed stone required (26.3).

The new formula for converting cubic yards to tons is:

Cubic yards multiplied by 1.35 equals tons of crushed stone required.

In our example, multiply 26.3 times 1.35. This gives you 35.5 tons of crushed stone, as shown in Figure 6-1.

You can see how factor tables can make your estimates easier, quicker and more accurate. Any time you can shorten your mathematical computations, you're reducing the chance of error.

Concrete Block

Figure 6-2 is a factor table for concrete block. Most foundations will use a combination of 12'' x 8'' x

Thickness	Factor	Foundation area (square feet)		Cubic yards
3"	.00926	x		=
3½"	.01080	x		=
4"	.01235	x	2,128.75	= 26.3
4½"	.01389	x	928	= 12.89
5"	.01543	x		=
5½"	.01698	x		=
6"	.01852	x		=

Cubic yards crushed stone multiplied by 1.35 = tons

Cu. yds. crushed stone 26.3 x 1.35 = 35.5 tons crushed stone
(ORDER 36 TONS)

Factors for Crushed Stone
Figure 6-1

16", 8" x 8" x 16" and 4" x 8" x 16" blocks in the foundation wall. Use the factors in Figure 6-2 to compute the necessary quantity of each of these blocks.

Some builders prefer to use the square foot method of estimating concrete block requirements. This method is also shown in Figure 6-2. Use the square foot method only when blocks are 8" in height and 16" in length.

When you use the square foot method, be sure to compute the 8" x 8" x 16" blocks separately from the 12" x 8" x 16" blocks. *And don't use the 1.125 factor to compute any quantity of 4" x 8" x 16" blocks.*

Figure 6-2 shows a sample foundation wall with a perimeter measuring 131'4" (131.33 linear feet). How many concrete blocks do we need for the wall? There are 8 courses of 12" x 8" x 16" blocks, 4 courses of 8" x 8" x 16" blocks, and 1 course of 4" x 8" x 16" solid blocks for the cap. Using the factors and calculations shown in Figure 6-2, the total number of blocks will be:

12" x 8" x 16"	788
8" x 8" x 16"	394
4" x 8" x 16"	99
Total	1,281 blocks

The wall area for the 12" x 8" x 16" blocks is 700.43 SF (131.33' times 5.33333'). If you're computing block requirements by the square foot method, enter this number on your worksheet, as shown in Figure 6-2. Multiply the 1.125 factor times 700.43 SF. We'll need 787.98 or 788 blocks. That's the same as the total we got by using the 0.75000 factor. Both methods are fast and accurate. Use the method that seems easier to you.

A stepped footing can reduce or increase the number of blocks you'll need for a foundation wall. Be sure to allow for this in your estimate.

Also remember that some block will be broken during construction. Breakage varies from job to job and from one mason to another. Record the breakage and waste that occurs on each job. Then compute the average loss for each type of job. Add this to your quantity when ordering. An allowance of 4% to 8% is usually a safe damage allowance.

Masonry Cement and Sand
Figure 6-3 will help you compute the bags of cement and tons of sand required for the job. There's a high percentage of waste in mortar. It can run as high as 25% or more. Again, cost records from your previous jobs will give you an idea of how much to allow for waste. The size of the mortar joint and the size of the block also affect the amount of waste.

Factor	Foundation length (linear feet)	Number of courses	Number of blocks
.75000	x *131.33' (131'4")* x	*8*	= *787.98 (or 788)*

Computing number of blocks (12" x 8" x 16")

Factor	Foundation length (linear feet)	Number of courses	Number of blocks
.75000	x *131.33' (131'4")* x	*4*	= *393.99 (or 394)*

Computing number of blocks (8" x 8" x 16")

Factor	Foundation length (linear feet)	Number of courses	Number of blocks
.75000	x *131.33' (131'4")* x	*1*	= *98.5 (or 99)*

Computing number of blocks (4" x 8" x 16")

Factor	Foundation area (square feet)	Number of blocks
1.125	x *700.43*	= *787.98 (or 788)*

Computing number of blocks by square foot method (blocks 8" in height x 16" in length)

Factors for Concrete Block
Figure 6-2

The factors in Figure 6-3 will work for nearly all block sizes. These factors include an allowance for waste and for variation in the size of mortar joints. Look at Figure 6-3 A. The first factor (0.02400) allows for limited waste. If you want a waste allowance of 25%, use the 0.03000 factor.

For the job described in Figure 6-2, we'll need to lay 1,281 concrete blocks. To find the number of bags of cement required, multiply the factor in Figure 6-3 A (we'll use 0.02400) times the number of blocks (1,281). We'll need 31 bags of cement. To find the number of tons of sand required, multiply the factor in Figure 6-3 B (0.00300) by the number of blocks (1,281). We'll need 4 tons of sand.

Parging

Basements and habitable spaces below grade must be protected against moisture. Parge foundation walls with 1/2" of masonry mortar. Then apply a bituminous coating. Install footing drain tile around the foundation. This will keep any standing water from penetrating the foundation wall. The waterproofing can be done any time after the foundation blocks have been laid and before backfilling around the foundation begins. Let's look at each of the materials you'll use for parging.

Mortar— Figure 6-4 is a worksheet for estimating the bags of masonry cement and tons of sand required for 1/2-inch parging. Be sure to enter a waste allowance in the space allotted for it.

Factor	Number of blocks	Bags of masonry cement
.02400	x *1,281*	= *30.74 (or 31)*
.03000 (25% waste)	x	=

A Bags of masonry cement

Factor	Number of blocks	Tons of sand
.00300	x *1,281*	= *3.84 (or 4)*

B Tons of sand

Factors for Masonry Cement and Sand
Figure 6-3

Factor	Foundation area (square feet)	Bags of masonry cement
.01848	x $\quad$ *902*	= *16.67*
Bags of masonry cement		*16.67*
Add for waste (*5* %)		*.83*
Total		*17.50 or 18*

A Bags of masonry cement

Factor	Foundation area (square feet)	Tons of sand
.00231	x $\quad$ *902*	= *2.084*
Tons of sand		*2.084*
Add for waste (*5* %)		*.104*
Total		*2.188*

B Tons of sand

**Factors for Parging Cement and Sand
Figure 6-4**

Figure 6-5 is a weekly time sheet and daily log for the week ending March 12 on the Brown job. The daily log shows that 1/2-inch parging is required for 902 SF of foundation wall. Here's how to use the factor table shown in Figure 6-4 A to compute the cement required for parging. Multiply the factor (0.01848) by the square feet of foundation area (902). This gives us 16.67 bags of masonry cement. Add a waste allowance of 5% for a total of 17.5, or 18 bags.

Use Figure 6-4 B to compute the sand required for the parging. Multiply the factor (0.00231) by the square feet of foundation area (902). This comes to 2.084 tons. Add a waste allowance of 5%. This makes a total of 2.188 tons of sand. Include this quantity of sand in your order for any other sand you'll need for the job. This will help give you a full truckload.

Bituminous coating— You can normally estimate the material for bituminous coating at 1 gallon per 100 SF of wall area. Use this figure unless the manufacturer makes some other recommendation.

Drain tile— Figure 6-6 is a worksheet for computing the cubic yards and tons of crushed stone required for foundation drain tile 18" wide and 12" deep. Assume that 4" drain tile is surrounded with crushed stone. Multiply the factor (0.05232) by the linear feet of foundation drain tile. This gives you the cubic yards of crushed stone. Multiply the number of cubic yards by 1.35 to get the tons of crushed stone required.

The calculation shown in Figure 6-6 is for the drain tile on the Brown job. Multiply the factor (0.05232) by the linear feet (106). This gives you 5.55 cubic yards of crushed stone. Then multiply 5.55 by 1.35. This gives you 7.49 tons of crushed stone. You'll probably order 8 tons of crushed stone for the job.

Foundation Labor

Estimating manhours is never easy. But if you keep accurate labor records on each phase of the job, you'll have an excellent guide to follow on your next estimate.

Let's look at a quick, accurate way to make labor estimates for foundation work. We'll cover all the important foundation wall components: crushed stone, concrete block, and parging.

Crushed Stone

When you place crushed stone under a concrete basement floor, spread the stone before the walls are set. This will save you time and money. The truck will be able to move in close to the foundation area. And the masons and carpenters will have a safer, drier working area. If you don't spread the crushed stone before the foundation walls go up, you'll have to haul it to the basement in wheelbarrows later.

Once the crushed stone is delivered to the basement area, additional labor is required to spread it evenly and bring it to the correct elevation. Look at Figure 6-7. This is a weekly time sheet and daily log for the week ending September 4 on the Brown job. The daily log tells us that on Monday, three workers spent 13 manhours laying out house corners for the masons and 8 manhours spreading gravel in the basement. On Tuesday, it took 4 manhours to finish spreading the stone. A total of 12 manhours were required to spread the 18 tons of stone. Only 4" of stone were required. But 4½" of crushed stone were spread. It's hard to spread exactly 4", and you should never spread less than that on a surface that's been machine-graded.

Weekly Time Sheet

For period ending 3-12-XX   *Brown* job

	Name	Exemptions	Days MARCH						Rate	Hours worked		Total earnings
			7 M	8 T	9 W	10 T	11 F	12 S		Reg.	Over-time	
1	O.L. White		8	8	6½	8	8	X		38½		
2	A.L. King		X	8	8	8	8	X		32		
3	T.E. King		X	X	8	8	8	X		24		
4	Holston Drywall, Inc.		X	X	X	✓	X	X		—		
5												
6												
7												
8												
9												
10												
11												
12												
13												
14												
15												
16												
17												
18												
19												
20												

Daily Log

Monday Started waterproofing foundation (8 manhours for parging)
Tuesday Finished 902 S.F. parging (8 manhours). Started bituminous coating (8 manhours)
Wednesday Finished 902 S.F. bituminous coating (6½ manhours). Drain tile around basement area (16 manhours)
Thursday Drywall. Outside trim.   106'
Friday Outside trim.
Saturday XXXX

Weekly Time Sheet
Figure 6-5

Factor	Foundation tile (linear feet)	Cubic yards crushed stone
.05232	x _106_	= _5.55_

Cubic yards crushed stone multiplied by 1.35 = tons

Cu. yds. crushed stone _5.55_ x 1.35 = _7.49_ tons crushed stone (order 8 tons)

Factors for Drain Tile
Figure 6-6

Figure 6-8 is a foundation labor worksheet. To compute the manhours required for spreading crushed stone under a basement area, use the formula shown in Figure 6-8 A. Here's how:

Total the foundation work manhours on your weekly time sheets. Enter the total for crushed stone labor in the first column of Figure 6-8 A. In the next column, enter the number of tons of crushed stone required for the job. Our weekly time sheets for the Brown job (Figures 6-5 and 6-7) tell us that 12 manhours were required to spread and level 18 tons of crushed stone.

Now divide the total manhours by the number of tons of crushed stone. This will give you the manhour factor. In our example, the total manhours (12) divided by the number of tons of crushed stone (18) gives us a manhour factor of 0.66667.

Use this manhour factor as a basis for future estimates. To estimate the labor for spreading crushed stone on a similar job, just multiply the manhour factor from Figure 6-8 A (0.66667) by the estimated number of *tons* of crushed stone. If your estimate shows 26 tons of crushed stone, then your manhour estimate for this job will be 17.33, or 18 manhours.

Now multiply the estimated manhours by the hourly rate of pay. This will give you the estimated labor cost for the job. For this example, 18 manhours times $25.00 equals an estimated labor cost of $450.00.

Concrete Block
Laying concrete block is normally subcontracted out to an experienced mason. He'll contract to do it either by the job or by the block. When the masonry work is done on contract, just enter the masonry subcontractor's name on your weekly time sheet, and use a check mark to indicate the days he worked. See Figure 6-7.

A masonry subcontractor will hire his own crew. He carries all responsibility for payroll records, taxes and insurance. If you pay masons and helpers by the hour, you'll also have to pay the FICA tax, FUTA tax, withholding taxes, worker's compensation and liability insurance premiums. List their names and manhours on your weekly time sheets. Keep payroll records for each employee so you can file accurate quarterly and year-end reports.

Parging
Let's look at the manhours required for waterproofing the 902 SF foundation wall for the Brown job. We'll look at parging mortar, bituminous coating and drain tile.

Mortar— Look again at Figure 6-5. The daily log shows that the waterproofing of the foundation began on Monday, March 7. D.L. White worked 8 hours parging that day. On the following day, 8 manhours were required to finish parging. A total of 16 manhours were required to parge the wall.

To convert this information to a manhour factor, use the worksheet in Figure 6-8 B. Divide the total manhours (16) by the foundation area in square feet (902). This gives us a manhour factor of 0.01774. Use this manhour factor as a basis for future parging estimates. Here's how:

Let's say you're estimating the labor for half-inch parging for 1,258 SF of foundation wall area. Multiply the factor (0.01774) by 1,258 SF. The manhour estimate for the work will be 22.32, or 23 manhours.

Now multiply the number of manhours (23) by the hourly rate of pay ($25.00). Your labor cost estimate for the job is $575.00.

Bituminous coating— Figure 6-5 indicates that work on the bituminous coating began on Tuesday, March 8. Workers spent 8 manhours on bituminous coating that day. On the following day, workers spent 6½ manhours finishing up the job. A total of 14½ manhours were required to apply bituminous coating on 902 SF of foundation wall area.

To convert this information to a manhour factor, use the worksheet in Figure 6-8 C. Divide the total manhours (14.5) by the foundation area in square feet (902). This gives us a manhour factor of 0.01608. Use this factor as a basis for future bituminous coating estimates.

If you're estimating labor for the bituminous coating on the job with 1,258 SF of foundation wall area, multiply the factor (0.01608) by 1,258 SF. The manhour estimate for the work will be 20.23, or 21 manhours.

Now multiply the number of manhours (21) by the hourly wage rate ($25.00). Your estimated labor cost for this job is $525.00.

Weekly Time Sheet

Page____of____pages

For period ending ___9-4-XX___   ___Brown___ job

	Name	Exemptions	Days Aug. SEPT						Rate	Hours worked		Total earnings
			30 M	31 T	1 W	2 T	3 F	4 S		Reg.	Over-time	
1	D.L. White		7	X	X	X	X	X		7		
2	J.E. King		7	2	X	X	X	X		9		
3	W.R. Farlow		7	2	X	X	X	X		9		
4	W.C. Martin, Masonry Sub.		X	X	✓	X	✓	X		—		
5												
6												
7												
8												
9												
10												
11												
12												
13												
14												
15												
16												
17												
18												
19												
20												

Daily Log

Monday LAID OUT HOUSE CORNERS (13 MANHOURS). GRAVEL IN BASEMENT (8 MANHOURS).

Tuesday FINISHED GRAVEL IN BASEMENT AREA (4 MANHOURS). Total 18 tons of GRAVEL.

Wednesday STARTED laying blocks for foundations.

Thursday RAIN.

Friday Blocks for foundation.

Saturday X X X

Weekly Time Sheet
Figure 6-7

Total manhours	÷	Tons of crushed stone	=	Manhour factor
12	÷	*18*	=	*.66667*

A Spreading crushed stone in basement area

Total manhours	÷	Foundation area (square feet)	=	Manhour factor
14.5	÷	*902*	=	*.01608*

C Brushing on bituminous coating

Total manhours	÷	Foundation area (square feet)	=	Manhour factor
16	÷	*902*	=	*.01774*

B Parging foundation walls

Total manhours	÷	Foundation tile (linear feet)	=	Manhour factor
16	÷	*106*	=	*.15094*

D Installing drain tile

Foundation Labor Worksheet
Figure 6-8

Drain tile— The daily log in Figure 6-5 shows that 16 manhours were spent installing 106 linear feet of drain tile around the basement foundation. Record this information on the worksheet shown in Figure 6-8 D. Divide the total manhours (16) by the linear feet of the drain tile (106) to get the manhour factor (0.15094).

If you're estimating the labor on a job requiring 126 linear feet of drain tile, just multiply the factor (0.15094) by the linear feet (126) to get your manhour estimate. This job will require about 19 manhours.

Now multiply the number of manhours by the hourly rate of pay ($25.00) to get the estimated cost of labor for the job.

In this chapter, we've seen how to keep accurate foundation cost records and make quick, accurate estimates for foundation materials and labor. In the next chapter, we'll look at cost records for the floor system.

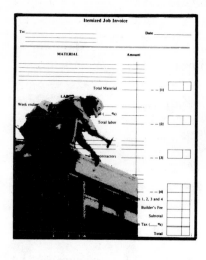

Chapter 7

Floor System Records

If you're putting up a factory-built house, you'll want to have the floor system in place before the other house components arrive at the job site. This saves time. You can unload the wall panels from the truck and take them directly to the correct position on the subfloor.

If the floor system and wall system are delivered at the same time, you can assemble the floor while the wall system waits on the truck. But remember, there's usually an extra charge if the truck isn't unloaded within a specified time limit.

In this chapter, you'll use your cost records to make fast, accurate floor system estimates. We'll take a detailed look at floor system materials and labor. And we'll look at a sample floor-system estimate.

Floor System Materials
The most commonly-used floor system materials are board lumber, plywood panels, and nails. Let's look at the easy way to estimate quantities and costs for each of these materials.

Board Lumber
Lumber is commonly measured in board feet or in linear feet. A board foot is the amount of lumber that will fill a space 1' long, 1' wide and 1" thick. Most lumber is priced by the thousand board feet (MBF). But hardwood and finish lumber are often priced by the linear foot.

Lumber measurements are based on ''nominal'' or name dimensions, not actual dimensions. For example, a nominal 2 x 12 that's 10 feet long contains 20 board feet of lumber.

Sometimes you'll need to buy lumber by the linear foot even though it's priced by the MBF. If so, you'll have to convert linear feet to board feet. The conventional formula for converting linear feet to board feet is:

Multiply the nominal dimensions by the length (linear feet) and divide by 12.

For example, let's assume that you placed an order for 380 linear feet of 1 x 4 S4S. How many board feet is that?

$$\frac{1 \times 4 \times 380 \text{ linear feet}}{12} = 126.67 \text{ board feet}$$

We can simplify this computation by using the factor tables shown in Figures 7-1A and 7-1B. Figure 7-1A is a list of factors for converting linear feet to board feet. Figure 7-1B shows how to use the factors. Look at the first column in Figure 7-1B. Find the nominal dimensions of the lumber in our example (1 x 4). In the next column, enter the number of linear feet (380). Multiply the number of linear feet by the factor for 1 x 4 lumber (0.33333) to get the number of board feet (126.67).

Lumber size (thickness and width)	Factors
1'' x 4''	.33333
1'' x 6''	.50000
1'' x 8''	.66667
1'' x 10''	.83333
1'' x 12''	1.00000
2'' x 4''	.66667
2'' x 6''	1.00000
2'' x 8''	1.33333
2'' x 10''	1.66667
2'' x 12''	2.00000

Factors for Converting Linear Feet to Board Feet
Figure 7-1A

Our new formula for converting linear feet to board feet is:

Multiply the linear feet by the factor to get the number of board feet.

Use the factors in Figure 7-1 to calculate the number of board feet in different lumber sizes. For a 2 x 4 that's 8' long, multiply the number of linear feet (8) by the factor for 2 x 4 lumber (0.66667) to get the number of board feet (5.33). (See Figure 7-1B.)

What if your measurements are in board feet, but the lumber is priced in linear feet? The conventional formula for converting board feet to linear feet is:

Multiply the board feet by 12 and divide by the nominal dimensions.

For example, how many linear feet are there in 1,280 board feet of 2 x 4 lumber?

$$\frac{1280 \text{ board feet x } 12}{2 \text{ x } 4} = 1{,}920 \text{ linear feet}$$

Now let's do it the easy way. Figure 7-2A is a list of factors for converting board feet to linear feet. Figure 7-2B shows how to use the factors. Look at the first column in Figure 7-2B. Find the nominal dimensions of the lumber in our example (2 x 4). In the next column enter the number of board feet (1,280). Now multiply the number of board feet by the factor for 2 x 4 lumber (1.50) to get the number of linear feet (1,920).

Lumber size (thickness and width)	Linear feet	x	Factor	=	Board feet
1'' x 4''	380	x	.33333	=	126.67
1'' x 6''		x	.50000	=	
1'' x 8''		x	.66667	=	
1'' x 10''		x	.83333	=	
1'' x 12''		x	1.00000	=	
2'' x 4''	8	x	.66667	=	5.33
2'' x 6''	8	x	1.00000	=	8.00
2'' x 8''		x	1.33333	=	
2'' x 10''	14	x	1.66667	=	23.33
2'' x 12''		x	2.00000	=	

Worksheet for Converting Linear Feet to Board Feet
Figure 7-1B

Our new formula for converting board feet to linear feet is:

Multiply board feet times the factor to get the number of linear feet.

Now let's say that lumber is priced by the piece, and you want to know how much it costs per MBF. For example, 2 x 10 joists 16' long are priced at $8.80 each. What is the cost per MBF? The conventional solution would be:

Convert linear feet to board feet:

$$\frac{2 \text{ x } 10 \text{ x } 16}{12} = 26.66667 \text{ board feet}$$

Divide price per piece by number of board feet:

$$\frac{\$8.80}{26.66667} = \$.33 \text{ cost per board foot}$$

Multiply cost per board foot times 1,000:
$.33 x 1,000 = $330.00 cost per MBF

Lumber size (thickness and width)	Factors
1" x 4"	3.00000
1" x 6"	2.00000
1" x 8"	1.50000
1" x 10"	1.20000
1" x 12"	1.00000
2" x 4"	1.50000
2" x 6"	1.00000
2" x 8"	.75000
2" x 10"	.60000
2" x 12"	.50000

Factors for Converting Board Feet to Linear Feet
Figure 7-2A

Now let's use the factor table shown in Figure 7-3 to find the price per MBF the easy way. Look down the first column until you find the lumber dimensions in our example (2 x 10 x 16). Read across from left to right. In the next column enter the price per piece ($8.80). Multiply the piece price by the factor (37.50) to get the price per MBF ($330.00).

Our new formula for calculating the price per MBF is:

Multiply the piece price by the factor to get the price per MBF.

Many lumber dealers sell their framing lumber by the piece rather than by the MBF. They do this to simplify computing the price on small purchases. Piece pricing isn't practical for random lengths of lumber. But it works well on any sale where the lumber is cut to uniform lengths.

Here's an example. Let's say 2 x 4 x 8 studs are priced at $1.65 each. Joists measuring 2 x 6 x 14 cost $4.41 each, and 2 x 6 x 16 joists go for $5.04 each. Let's assume you need them in quantities of 400, 66 and 40. Here's how you'd calculate the total cost:

400 -	2 x 4 x 8 @ $1.65	=	$660.00
66 -	2 x 6 x 14 @ $4.41	=	$291.06
40 -	2 x 6 x 16 @ $5.04	=	$201.60
	Total		$1,152.66

Lumber size (thickness and width)	Board feet	x	Factor	=	Linear feet
1" x 4"	126.67	x	3.00	=	380
1" x 6"	140	x	2.00	=	280
1" x 8"		x	1.50	=	
1" x 10"		x	1.20	=	
1" x 12"		x	1.00	=	
2" x 4"	1,280	x	1.50	=	1,920
2" x 6"		x	1.00	=	
2" x 8"		x	.75	=	
2" x 10"	23.33	x	.60	=	14
2" x 12"		x	.50	=	

Worksheet for Converting Board Feet to Linear Feet
Figure 7-2B

If lumber is priced by the MBF, and you want to buy it by the piece, the cost calculation is more difficult. Assuming again that you're purchasing this lumber in quantities of 400, 66 and 40, the conventional math looks like this:

$$\frac{400 \times 2 \times 4 \times 8}{12 \times 1,000} \times \$309.38 = \$660.01$$

$$\frac{66 \times 2 \times 6 \times 14}{12 \times 1,000} \times \$315.00 = \$291.06$$

$$\frac{40 \times 2 \times 6 \times 16}{12 \times 1,000} \times \$315.00 = \$201.60$$

Now look at Figure 7-4. Use this factor table to simplify your calculation when lumber is priced by the MBF and you want to buy it by the piece. Here's how:

Look down the first column until you find the lumber dimensions in our example. Let's look first at the 2 x 4 x 8 studs. In the next column enter the number of pieces (400). Multiply the number of pieces by the factor for 2 x 4 x 8 lumber (0.0053333)

Lumber size	Price per piece	x	Factor	=	Price per MBF
2 x 4 x 8	$1.65	x	187.50	=	$309.38
2 x 4 x 10	2.07	x	150.00	=	310.50
2 x 4 x 12	2.48	x	125.00	=	310.00
2 x 4 x 14	2.89	x	107.14290	=	309.64
2 x 4 x 16	3.36	x	93.74997	=	315.00
2 x 4 x 18	3.78	x	83.33333	=	315.00
2 x 4 x 20	4.20	x	75.00	=	315.00
2 x 6 x 8	2.52	x	125.00	=	315.00
2 x 6 x 10	3.15	x	100.00	=	315.00
2 x 6 x 12	3.78	x	83.33333	=	315.00
2 x 6 x 14	4.41	x	71.42857	=	315.00
2 x 6 x 16	5.04	x	62.50	=	315.00
2 x 6 x 18	5.76	x	55.55556	=	320.00
2 x 6 x 20	6.40	x	50.00	=	320.00
2 x 8 x 8	3.31	x	93.74997	=	310.31
2 x 8 x 10	4.13	x	75.00	=	309.75
2 x 8 x 12	4.96	x	62.50	=	310.00

**Factors for converting price per piece
into price per MBF
Figure 7-3**

Lumber size	Price per piece	x	Factor	=	Price per MBF
2 x 8 x 14	$5.79	x	53.57142	=	$310.18
2 x 8 x 16	6.61	x	46.87501	=	309.84
2 x 8 x 18	7.68	x	41.66667	=	320.00
2 x 8 x 20	8.53	x	37.50	=	319.88
2 x 10 x 8	4.40	x	75.00	=	330.00
2 x 10 x 10	5.50	x	60.00	=	330.00
2 x 10 x 12	6.60	x	50.00	=	330.00
2 x 10 x 14	7.70	x	42.85715	=	330.00
2 x 10 x 16	8.80	x	37.50	=	330.00
2 x 10 x 18	10.20	x	33.33333	=	340.00
2 x 10 x 20	11.33	x	30.00	=	339.90
2 x 12 x 8	6.72	x	62.50	=	420.00
2 x 12 x 10	8.40	x	50.00	=	420.00
2 x 12 x 12	10.08	x	41.66667	=	420.00
2 x 12 x 14	11.76	x	35.71429	=	420.00
2 x 12 x 16	13.44	x	31.25	=	420.00
2 x 12 x 18	15.12	x	27.77778	=	420.00
2 x 12 x 20	16.80	x	25.00	=	420.00

**Factors for converting price per piece
into price per MBF
Figure 7-3 (continued)**

times the price per MBF ($309.38). This gives the total lumber cost of $660.01.

Now let's look at the 2 x 6 x 14 joists. Find the lumber dimensions in the first column. In the next column enter the number of pieces (66). Multiply the number of pieces by the factor for 2 x 6 x 14 lumber (0.0140) times the price per MBF ($315.00). The total lumber cost is $291.06.

Use the same procedure for the 2 x 6 x 16 joists. Find the dimensions, and enter the number of pieces. Then multiply the number of pieces by the factor times the price per MBF. The total lumber cost is $201.60.

Plywood Panels
The conventional formula for estimating plywood floor sheathing requirements is:

Square feet of floor area divided by the square feet in one piece of plywood. Round fractions up to the next whole number.

Lumber size	Number of pieces	x	Factor	x	Price per MBF	=	Cost of lumber
2 x 4 x 8	400	x	.0053333	x	$309.38	=	660.01
2 x 4 x 10	35	x	.0066667	x	310.50	=	72.45
2 x 4 x 12	50	x	.0080	x	310.00	=	124.00
2 x 4 x 14	20	x	.0093330	x	309.64	=	57.80
2 x 4 x 16	12	x	.0106667	x	315.00	=	40.32
2 x 4 x 18	14	x	.0120	x	315.00	=	52.92
2 x 4 x 20	16	x	.0133333	x	315.00	=	67.20
2 x 6 x 8	1	x	.0080	x	315.00	=	2.52
2 x 6 x 10	20	x	.0100	x	315.00	=	63.00
2 x 6 x 12	5	x	.0120	x	315.00	=	18.90
2 x 6 x 14	66	x	.0140	x	315.00	=	291.06
2 x 6 x 16	40	x	.0160	x	315.00	=	201.60
2 x 6 x 18	40	x	.0180	x	320.00	=	230.40
2 x 6 x 20	4	x	.0200	x	320.00	=	25.60
2 x 8 x 8	1	x	.0106667	x	310.31	=	3.31
2 x 8 x 10	8	x	.0133333	x	309.75	=	33.04
2 x 8 x 12	28	x	.0160	x	310.00	=	138.88

Lumber size	Number of pieces	x	Factor	x	Price per MBF	=	Cost of lumber
2 x 8 x 14	6	x	.01866667	x	310.18	=	34.74
2 x 8 x 16	115	x	.0213333	x	309.84	=	760.14
2 x 8 x 18	15	x	.0240	x	320.00	=	115.20
2 x 8 x 20	12	x	.0266667	x	319.88	=	102.36
2 x 10 x 8	1	x	.0133333	x	330.00	=	4.40
2 x 10 x 10	25	x	.0166667	x	330.00	=	137.50
2 x 10 x 12	16	x	.0200	x	330.00	=	105.60
2 x 10 x 14	114	x	.0233333	x	330.00	=	877.80
2 x 10 x 16	108	x	.0266667	x	330.00	=	950.40
2 x 10 x 18	4	x	.0300	x	340.00	=	40.80
2 x 10 x 20	8	x	.0333333	x	339.90	=	90.64
2 x 12 x 8	1	x	.0160	x	420.00	=	6.72
2 x 12 x 10	12	x	.0200	x	420.00	=	100.80
2 x 12 x 12	44	x	.0240	x	420.00	=	443.52
2 x 12 x 14	2	x	.0280	x	420.00	=	23.52
2 x 12 x 16	88	x	.0320	x	420.00	=	1182.72
2 x 12 x 18	4	x	.0360	x	420.00	=	60.48
2 x 12 x 20	5	x	.0400	x	420.00	=	84.00

Factors for converting price per MBF into cost for total number of pieces
Figure 7-4

For example, let's assume there are 1,814 SF of floor area to be covered with 4 x 8 (32 SF) panels. How many plywood panels will you need?

$$\frac{1,814 \text{ SF}}{32} = 56.69 \text{ (57 pieces)}$$

Now let's do it the easy way. Look at Figure 7-5 A. This table shows the factor for 4 x 8 plywood panels. Look down the first column and find the floor area in our example (1,814 SF). Multiply the floor area by the factor (0.03125) to get the number of plywood panels required. The total comes to 56.69 panels. Round this up to 57 panels. When you're using this table, keep in mind that *the factor does not include a waste allowance.*

Our new formula for calculating plywood panel requirements is:

Multiply square feet of floor area by the factor. Round fractions up to the next whole number. Factor does not include waste allowance.

Figure 7-5 B shows the factor for 4 x 12 plywood

panels.

Nails

Nails for the framing, sheathing, floor and roof decks are normally purchased in 50- or 100-lb. quantities. Most lumber dealers also sell nails by the pound. But the pound price is usually higher than the keg or 50-lb. price. Most builders buy framing nails by the carton or keg.

Provide extra nails for every job. Some of the nails you order will be needed for miscellaneous applications, such as forms, braces and scaffolding. The factors shown in Figure 7-6 allow for 15% waste.

Spacing of the floor joists (either 16'' o.c. or 24'' o.c.) will change the quantity of nails needed. For example, a plywood floor deck with joists 16'' o.c. will require 15 lbs. of 8d common nails per 1,000 SF. If the floor joists are spaced 24'' o.c., you'll need only 11.5 lbs. of 8d common nails per 1,000 SF. The factors shown in Figure 7-6 are for floor joists spaced 16'' o.c.

Area (square feet)	x	Factor	=	Number of pieces
1,040	x	.03125	=	32.50 (or 33)
1,280	x	.03125	=	40
1,450	x	.03125	=	45.31 (or 46)
1,676	x	.03125	=	52.38 (or 53)
1,814	x	.03125	=	56.69 (or 57)

A 4' x 8' plywood

Area (square feet)	x	Factor	=	Number of pieces
1,040	x	.02083	=	21.66 (or 22)
1,280	x	.02083	=	26.66 (or 27)
1,450	x	.02083	=	30.20 (or 31)
1,676	x	.02083	=	34.91 (or 35)
1,814	x	.02083	=	37.79 (or 38)

B 4' x 12' plywood

Factors for plywood panels
Figure 7-5

Total board feet	x	Factor	=	Pounds nails
3,708	x	.01000	=	37.08 (or 37)

Allow 10 lbs. 16d common nails per 1,000 board feet
Floor joists and headers

Total board feet	x	Factor	=	Pounds nails
3,708	x	.00250	=	9.27 (or 10)

Allow 2.5 lbs. 8d common nails per 1,000 board feet
Toenailing joist headers and floor joists

Total linear feet	x	Factor	=	Pounds nails
300	x	.03500	=	10.50 (or 11)

Allow 3.5 lbs. 8d common nails per 100 linear feet
Wood cross-bridging

Factors for nails
Figure 7-6

Total square feet	x	Factor	=	Pounds nails
	x	.00800	=	
1,814				14.51 (or 15)

Allow 8 lbs. 6d threaded nails, per 1,000 square feet

Plywood floor deck - 6d nails

Total square feet	x	Factor	=	Pounds nails
	x	.01500	=	
1,814				27.21 (or 27)

Allow 15 lbs. 8d common nails per 1,000 square feet

Plywood floor deck - 8d nails

Total square feet	x	Factor	=	Pounds nails
	x	.02700	=	
1,040				28.08 (or 28)

Allow 27 lbs. 7d threaded nails per 1,000 square feet

Board floor deck - 7d nails

Total square feet	x	Factor	=	Pounds nails
	x	.04000	=	
1,040				41.60 (or 42)

Allow 40 lbs. 8d common nails per 1,000 square feet

Board floor deck - 8d nails

Factors for nails
Figure 7-6 (continued)

Floor System Labor

The manhours required to install a floor system will vary with the size and design of the house, the weather, and the efficiency of your crew. Accurate labor estimates are almost impossible without your records from previous jobs. Use your weekly time sheets and daily logs as a guide.

Figures 7-7A, 7-7B and 7-7C are weekly time sheets for the Baker job. The Baker house is a large two-story house with a crawl space but no basement. The floor joists rest on a steel beam. The

first-floor system was installed before the factory-built house arrived on the job site.

Look at the daily logs. On Wednesday, June 30, workers spent 16 manhours installing the foundation sill plate. On Friday, July 2, they spent 32 manhours installing the steel beams and floor joists. Monday, July 5, was a holiday. On Tuesday, July 6, workers installed the bridging. They began installing the subfloor, but rain forced them to quit for the day. Only 14 manhours were recorded. On Wednesday, July 7, the crew spent 32 manhours finishing up the subfloor. A total of 94 manhours were required to install a first-floor system of 1,956 SF.

Let's use this information to find a manhour factor for estimating labor on future jobs. Figure 7-8 is a floor system labor worksheet. Look at Figure 7-8 A. Enter the total manhours (94) in the first column. Enter the floor area (1,956 SF) in the next column. Divide the manhours by the floor area to get the manhour factor (0.04806).

Use this manhour factor to estimate labor requirements for similar jobs. Here's the formula:

Multiply the manhour factor by the floor area (square feet) to get the total number of manhours required to do the job. Then multiply the total manhours by the hourly pay rate to get the estimated labor cost.

Use the same procedure to estimate the labor required for the second floor. Look at Figure 7-7C. The daily log shows that on Wednesday, July 14, workers spent 72 manhours installing the floor joists on the second floor. On Thursday, July 15, they worked 60 hours installing the subfloor. It took the workers 132 manhours to install a second-floor system of 1,696 SF.

Look at Figure 7-8 B. Enter the total manhours (132) in the first column. Enter the floor area (1,696 SF) in the next column. Divide the manhours by the floor area. Our manhour factor for the second floor is 0.07783.

Figures 7-9A and 7-9B are weekly time sheets for the Brown job. This is a two-story house with a basement. It has a special floor system that requires floor trusses. The house is factory-built, and the first-floor system is included in the house package.

The daily log in Figure 7-9A tells us that the factory-built house was delivered on Tuesday, September 14, along with the first-floor system. Workers spent 48 manhours unloading the truck and installing most of the first-floor system. On Wednesday, September 15, they spent 33

Weekly Time Sheet

For period ending 7-3-XX   Baker _____ job

	Name	Exemptions	June 28 M	June 29 T	June 30 W	July 1 T	July 2 F	July 3 S	Rate	Reg.	Over-time	Total earnings
			Days							**Hours worked**		
1	D.L. White		7½	5½	8	8	8	X		37		
2	W.C. Martin, Masonry Contractor		✓	✓	X	X	X	X		—		
3	A.L. King		X	X	8	8	8	X		24		
4	J.E. King		X	X	8	8	8	X		24		
5	F.N. Neal		X	X	8	8	8	X		24		
6												
7												
8												
9												
10												
11												
12												
13												
14												
15												
16												
17												
18												
19												
20												

Daily Log

Monday _Foundation blocks--coordinating work with masons (7½ Manhours)_
Tuesday _Finished Foundation blocks -- coordinating work (5½ Manhours)_
Wednesday _Foundation sill plate (16 Manhours)-- Fill dirt in garage area (16 Manhours)_
Thursday _Waterproofing Foundation (16 Manhours)-- Fill dirt in porch (16 Manhours)_
Friday _Set steel beams and floor joists (32 Manhours)_
Saturday XXXXX

Weekly time sheet
Figure 7-7A

Weekly Time Sheet

For period ending _7-10-XX_   _Baker_ job

#	Name	Exemptions	5 M	6 T	7 W	8 T	9 F	10 S	Rate	Reg.	Over-time	Total earnings
			Days July							**Hours worked**		
1	D.L. White		X	3½	8	8	X	X		19½		
2	A.L. King		X	3½	8	8	X	X		19½		
3	J.E. King		X	3½	8	8	X	X		19½		
4	F.N. Neal		X	3½	8	8	X	X		19½		
5												
6												
7												
8												
9												
10												
11												
12												
13												
14												
15												
16												
17												
18												
19												
20												

Daily Log

Monday _Holiday (July 4th)_
Tuesday _Bridging -- started subfloor (14 Manhours) -- Rain_
Wednesday _Finished subfloor (32 Manhours) 1956 Sq. Ft. on 1st floor_
Thursday _Fill dirt in garage area (32 Manhours)_
Friday _XXXXX -- Waiting on delivery of Factory-built house_
Saturday _XXXXX_

Weekly time sheet
Figure 7-7B

Weekly Time Sheet

For period ending 7-17-XX

Baker job

	Name	Exemptions	Days July 12 M	13 T	14 W	15 T	16 F	17 S	Rate	Hours worked Reg.	Over-time	Total earnings
1	R.R. Lewis		8	8	8	X	X	X		24		
2	C.A. Lester		8	8	8	7½	8	X		39½		
3	D.L. White		8	8	8	7½	8	X		39½		
4	A.L. King		8	8	8	7½	8	X		39½		
5	J.E. King		8	8	8	7½	8	X		39½		
6	F.N. Neal		8	8	8	7½	8	X		39½		
7	W.R. Farlow		8	8	8	7½	8	X		39½		
8	R.C. Jones		X	4½	8	7½	8	X		28		
9	A.A. Norman		X	8	8	7½	X	X		23½		
10												
11												
12												
13												
14												
15												
16												
17												
18												
19												
20												

Daily Log

Monday Factory-built house arrived
Set o/s partitions on 1st floor (56 Manhours)
Tuesday Set i/s partitions on 1st floor (68½ Manhours)
Wednesday Set floor joists on 2nd floor (72 Manhours)
Thursday Subfloor on 2nd floor (60 Manhours) 4696 sq. ft. on 2nd floor
Friday Unloaded 2nd trailer (56 Manhours)
Saturday XXXXX

Weekly time sheet
Figure 7-7C

Total manhours	÷	Floor area (square feet)	=	Manhour factor
94	÷	1,956	=	.04806

A First floor (over crawl space)

Total manhours	÷	Floor area (square feet)	=	Manhour factor
132	÷	1,696	=	.07783

B Second floor (over crawl space)

Total manhours	÷	Floor area (square feet)	=	Manhour factor
81	÷	1,230	=	.06585

C First floor (over basement)

Total manhours	÷	Floor area (square feet)	=	Manhour factor
108	÷	1,040	=	.10385

D Second floor (over basement)

**Floor System Labor Worksheet
Figure 7-8**

manhours finishing the subfloor. A total of 81 manhours were required to install the first-floor system. The total floor area was 1,230 SF. This gives us a manhour factor of 0.06585 for the first floor. See Figure 7-8 C.

The daily log in Figure 7-9B shows that work on the second-floor system for the Brown job began on Tuesday, September 21. It required 48 manhours. On Wednesday, September 22, workers spent 40 manhours installing the floor joists and bridging. The subfloor was finished on Thursday, September 23. This work required 20 manhours. A total of 108 manhours were required to install the second-floor system. The total floor area was 1,040 SF. Our manhour factor for this work is 0.10385. See Figure 7-8 D.

Keep daily logs for all of your jobs. Convert the manhours to manhour factors. Use these factors to estimate labor costs on future jobs.

Floor System Estimate (Sample)

Now that we've learned the easy way to estimate floor system materials and labor, let's apply what we've learned. Let's estimate the materials and labor for the following job.

The first floor has 1,814 SF of floor area over a basement. A steel beam will be used. Here's how to estimate the material and labor costs for this job.

Materials

Let's estimate the board lumber, plywood panels and nails required for the job.

Board lumber— The lumber required for sill plates, floor joists and band joists will be:

5 - 2 x 6 x 12 (60 BF) @ $315.00 MBF

28 - 2 x 8 x 12 (448 BF) @ $310.00 MBF

16 - 2 x 10 x 12 (320 BF) @ $330.00 MBF

108 - 2 x 10 x 16 (2,880 BF) @ $330.00 MBF

Use the factors in Figure 7-4 to compute the cost of the framing lumber.

5 (2 x 6 x 12) x 0.0120	x $315.00	= $ 18.90
28 (2 x 8 x 12) x 0.0160	x $310.00	= $ 138.88
16 (2 x 10 x 12) x 0.0200	x $330.00	= $ 105.60
108 (2 x 10 x 16) x 0.0266667	x $330.00	= $ 950.40
Total		$1,213.78

The material for cross-bridging will be 1'' by 4''. It's priced at $12.80 per hundred linear feet. The multiplying factor is: 0.01000. We estimate that 300 linear feet are required. To find the cost, multiply the linear feet (300) by the factor (0.01000) times $12.80. This comes to $38.40.

Plywood panels— To find the number of pieces of 4 x 8 plywood, we'll use the factor in Figure 7-5 (0.03125). Remember that this factor doesn't include any allowance for waste.

Multiply the floor area (1,814 SF) by the factor (0.03125) to get the number of plywood panels. This comes to 56.69. Round up to 57 pieces.

Nails— Use the factor table in Figure 7-6 to estimate the number of pounds of nails required.

This job requires 3,708 BF of framing lumber. We'll use 16d common nails. The factor is 0.01000. Multiply the total board feet (3,708) by the factor (0.01000) to get the pounds of nails required (37.08, or 37).

Weekly Time Sheet

For period ending 9-18-XX   Brown ____job

	Name	Exemptions	Days September 13 M	14 T	15 W	16 T	17 F	18 S	Rate	Hours worked Reg.	Over-time	Total earnings
1	C.A. Lester		X	8	5½	3	8	X		24½		
2	D.L. White		X	8	5½	3	8	X		24½		
3	A.L. King		X	8	5½	2½	8	X		24		
4	J.E. King		X	8	5½	2½	8	X		24		
5	W.R. Farlow		X	8	5½	2½	X	X		16		
6	J.R. Davis		X	8	5½	2½	8	X		24		
7												
8												
9												
10												
11												
12												
13												
14												
15												
16												
17												
18												
19												
20												

Daily Log

Monday XXXXX --- Factory-built house due tomorrow
Factory-built house arrived with floor system for 1st floor
Tuesday Floor system on 1st floor (48 manhours)
Wednesday Finished subfloor on 1st floor (33 manhours) 1,230 sq. ft. on 1st floor
Thursday Unloaded second trailer -- Rain (16 manhours)
Friday Set o/s and i/s partitions on 1st floor (40 manhours)
Saturday XXXXX

Weekly time sheet
Figure 7-9A

Weekly Time Sheet

For period ending ___9-25-XX___   ___Brown___ job

	Name	Exemptions	Days September 20 M	21 T	22 W	23 T	24 F	25 S	Rate	Hours worked Reg.	Over-time	Total earnings
1	C.A.Lester		8	8	8	8	8	X		40		
2	D.L.White		8	8	8	8	8	X		40		
3	A.L.King		8	8	8	8	4	X		36		
4	W.R.Farlow		8	8	4	4	8	X		32		
5	J.R.Davis		8	8	8	8	8	X		40		
6	R.R.Lewis		X	8	4	4	8	X		24		
7												
8												
9												
10												
11												
12												
13												
14												
15												
16												
17												
18												
19												
20												

Daily Log

Monday Top plates on 1st floor partitions (40 Manhours)

Tuesday Set beams and started floor joists on 2nd floor (48 Manhours)

Wednesday Finished floor joists and bridging on 2nd floor (40 Manhours)

Thursday Finished subfloor (20 Manhours on 2nd floor) 1,040 sq. Ft. -- o/s partitions (20 Manhours)

Friday Finished o/s and i/s partitions on 2nd floor (36 Manhours): (Roof trusses)(8 Manhours)

Saturday XXXXX

Weekly time sheet
Figure 7-9B

For toenailing headers and floor joists, we'll use 8d common nails. The factor is 0.00250. Multiply the total board feet (3,708) by the factor (0.00250) to get the pounds of nails required (9.27, or 10).

For cross-bridging, we'll use 8d common nails. The factor is 0.03500. Multiply the total linear feet (300) by the factor (0.03500) to get the pounds of nails required (10.50, or 11).

For the plywood deck (with floor joists spaced 16'' o.c.), we'll use 6d threaded nails. The factor is 0.00800. Multiply the total square feet (1,814) by the factor (0.00800) to get the pounds of nails required (14.51, or 15).

Labor
Figure 7-8 C gives the manhour factor for a similar job with a first-floor system over a basement. Use the manhour factor from this job (0.06585) to estimate the labor requirements for your new job. Multiply the floor area (1,814 SF) by the factor (0.06585) to get the number of manhours required (119.45, or 120). Multiply the number of manhours (120) by the hourly pay rate ($25.00) to get the estimated labor cost for the job ($3,000.00).

In this chapter, we've seen how to use our cost record system to make fast, accurate floor system estimates. Next, we'll learn the easy way to estimate labor and materials for the wall system.

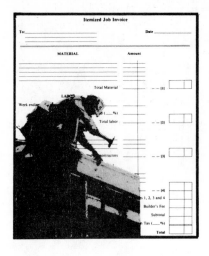

Chapter 8

Wall System Records

Most wall systems include the following components:

• Framing for the interior walls, including the sole and top plates, studs and headers

• Framing for the exterior walls, including the sole and top plates, studs and headers

• Sheathing for the exterior walls

• Nails for the framing lumber and wall sheathing

The cost of wall system materials and labor can vary considerably. If the wall sections arrive preassembled and with the wall sheathing already applied, your labor costs will be minimal. But your material costs will be high. If the workmen have to cut and assemble the framing lumber on site, your labor costs will increase. But your materials will cost less.

Which system should you use? The tables and worksheets in this chapter will help you decide which system is best for you. We'll use our cost records as a guide to easy, accurate estimates for wall system materials and labor.

Wall System Materials

Preassembled wall systems don't require many calculations for materials. The wall system will arrive with the major components already in place.

The manufacturer of the preassembled wall system will quote you a price based on your blueprints. The quoted price usually includes delivery but doesn't include sales tax. Be sure to add on the sales tax before submitting your bid.

If you're assembling the wall system on site, you'll need to make accurate estimates for framing lumber, sheathing and nails. Here's the easy way to do it:

Framing Lumber

Framing lumber for an on-site wall system includes: sole plates, top plates, studs, temporary braces, and headers. Let's take a look at each of these components.

Sole plates and top plates— Use your normal estimating procedures for sole plates, top plates, studs and temporary braces. Calculate the number of pieces required for each lumber size. Then use the factor table shown in Figure 8-1 to compute the cost of the lumber.

If you're using double top plates plus a sole plate, you'll need to find the lumber required for three plates. To do this, multiply the number of plates (3) by the total linear feet of exterior and interior walls. (Calculate the linear feet of exterior and interior walls for each floor, and then combine them to get the total linear feet.)

Let's try an example. If there are 860 linear feet of exterior and interior walls, multiply the linear feet of walls (860) by the number of plates (3) to get a total of 2,580 linear feet of lumber required for top and sole plates.

To get the number of pieces of lumber required for the plates, divide the total linear feet of lumber (2,580) by the length of the lumber being used. For example, if you're using 2 x 4 x 12's, divide the total linear feet (2,580) by the lumber length (12) to get a total of 215 pieces of 2 x 4 x 12 lumber required for the sole and top plates.

Now use the factor table in Figure 8-1 to compute the cost of the lumber. Look down the first column and find the lumber size. In our example, it's 2 x 4 x 12. In the next column, enter the number of pieces (215). Multiply the number of pieces by the factor (0.0080) to get 1.72. Multiply 1.72 by the

price per MBF ($310.00) to get a lumber cost of $533.20 for the top and sole plates.

Studs— It's easy to estimate the number of studs required for a straight wall that doesn't have any window openings, door openings or intersecting walls. But you'll often have to estimate stud requirements for walls with window openings, door openings, corner posts, T-posts, trimmer studs and blocking. Here's how.

Allow one stud for each linear foot of wall space. For example, if the total wall length comes to 860 linear feet, estimate 860 studs for the job.

Temporary braces— Most builders use 2 x 4 x 12's as temporary braces to keep the walls plumb and aligned until the upper top plate and ceiling

Lumber size	Number of pieces	x	Factor	x	Price per MBF	=	Cost of Lumber
2 x 4 x 8	722	x	.0053333	x	$310.00	=	$1,193.70
2 x 4 x 10	3	x	.0066667	x	310.00	=	6.20
2 x 4 x 12	215	x	.0080	x	310.00	=	533.20
2 x 4 x 14		x	.0093330	x		=	
2 x 4 x 16		x	.0106667	x		=	
2 x 4 x 18		x	.0120	x		=	
2 x 4 x 20		x	.0133333	x		=	
2 x 6 x 8	2	x	.0080	x	315.00	=	5.04
2 x 6 x 10	2	x	.0100	x	315.00	=	6.30
2 x 6 x 12	7	x	.0120	x	315.00	=	26.46
2 x 6 x 14		x	.0140	x		=	
2 x 6 x 16		x	.0160	x		=	
2 x 6 x 18		x	.0180	x		=	
2 x 6 x 20		x	.0200	x		=	
2 x 8 x 8	2	x	.0106667	x	310.00	=	6.61
2 x 8 x 10		x	.0133333	x		=	
2 x 8 x 12		x	.0160	x		=	

Factors for framing lumber
(priced by MBF)
Figure 8-1

Lumber size	Number of pieces	x	Factor	x	Price per MBF	=	Cost of Lumber
2 x 8 x 14		x	.01866667	x		=	
2 x 8 x 16		x	.0213333	x		=	
2 x 8 x 18		x	.0240	x		=	
2 x 8 x 20		x	.0266667	x		=	
2 x 10 x 8		x	.0133333	x		=	
2 x 10 x 10	4	x	.0166667	x	330.00	=	22.00
2 x 10 x 12		x	.0200	x		=	
2 x 10 x 14		x	.0233333	x		=	
2 x 10 x 16		x	.0266667	x		=	
2 x 10 x 18		x	.0300	x		=	
2 x 10 x 20		x	.0333333	x		=	
2 x 12 x 8		x	.0160	x		=	
2 x 12 x 10		x	.0200	x		=	
2 x 12 x 12	2	x	.0240	x	420.00	=	20.16
2 x 12 x 14		x	.0280	x		=	
2 x 12 x 16		x	.0320	x		=	
2 x 12 x 18		x	.0360	x		=	
2 x 12 x 20		x	.0400	x		=	

**Factors for framing lumber
(priced by MBF)
Figure 8-1 (continued)**

joists (or trusses) are installed. Allow about 20 pieces of 2 x 4 x 12 lumber for temporary braces in an average-size home.

Headers for interior door openings— Estimating the material you'll need for door and window headers can be time-consuming. The tables shown in Figures 8-2 and 8-3 will save you hours of work. They're tables for estimating interior and exterior headers.

Look at Figure 8-2. We'll use this table to estimate the header material required for the following example.

Let's say you have an interior door 6'0'' wide. You'll use two headers (placed on edge) for each opening. Look down the first column in Figure 8-2 until you find the door width in our example (6'0''). Read across from left to right. The table shows that we'll need two 2 x 8's for each opening.

The next column tells us that each header should be 6'5'' in length. Here's how this header length was calculated. We started with the door width (6'0'') and added to it:

1) Stud thickness— Each header should rest on one stud at each end of the opening. Each stud is 1½'' thick. This adds a total of 3'' to the header length.

2) Jamb thickness— Add another 2'' for the jamb thickness of the interior door frame.

Door width	Lumber size (two on edge)	Header length (each)	Estimate for each opening
2'0''	2 - 2 x 4	2'5''	1 - 2 x 4 x 8
2'4''	2 - 2 x 4	2'9''	1 - 2 x 4 x 8
2'6''	2 - 2 x 4	2'11''	1 - 2 x 4 x 8
2'8''	2 - 2 x 6	3'1''	1 - 2 x 6 x 8
3'0''	2 - 2 x 6	3'5''	1 - 2 x 6 x 8
4'0''	2 - 2 x 6	4'5''	1 - 2 x 6 x 10
5'0''	2 - 2 x 6	5'5''	1 - 2 x 6 x 12
6'0''	2 - 2 x 8	6'5''	1 - 2 x 8 x 14

Estimating headers for interior door openings
Figure 8-2

Rough stud opening	Lumber size (two on edge)	Header length (each)	Estimate for each opening
3'0''*	2 - 2 x 4	3'3''	1 - 2 x 4 x 8
4'0''*	2 - 2 x 6	4'3''	1 - 2 x 6 x 10
5'0''*	2 - 2 x 6	5'3''	1 - 2 x 6 x 12
6'0''*	2 - 2 x 8	6'3''	2 - 2 x 8 x 8
7'0''*	2 - 2 x 8	7'3''	2 - 2 x 8 x 8
8'0''**	2 - 2 x 10	8'6''	2 - 2 x 10 x 10
9'0''**	2 - 2 x 10	9'6''	2 - 2 x 10 x 10
10'0''**	2 - 2 x 12	10'6''	2 - 2 x 12 x 12
12'0''**	2 - 2 x 12	12'6''	2 - 2 x 12 x 14

*Headers bear on 1 - 2 x 4 on each end
**Triple studs at jamb opening. Headers bear on 2 - 2 x 4's on each end

Estimating headers for exterior door and window openings
Figure 8-3

In our example, the total header length equals the door width (6'0'') plus the stud thickness (3'') plus the jamb thickness (2''). This gives us a total header length of 6'5''.

There will be two headers for each opening. Multiplying 6'5'' by 2 gives us 12'10''. With this in mind, look at the last column in Figure 8-2. It shows that we'll need one 2 x 8 x 14 for each 6'0'' door width.

Headers for exterior door and window openings— Look at Figure 8-3. We'll use this table to estimate the header material required for the following example.

For rough stud openings of 10'0'', two 2 x 12 headers are required. Each header will be 10'6'' long. Notice that for all rough stud openings 8'0'' to 12'0'', headers must bear on two 2 x 4's at each end. For each 10' rough stud opening, you'll need two 2 x 12 x 12 headers.

Wall Sheathing

The most common exterior wall sheathing materials are fiberboard and plywood. Here's the easy way to estimate these materials for an on-site wall system:

First check the wall section of the blueprints to find the type and thickness of material required. Then calculate the total linear feet of exterior walls on each floor. Multiply the total linear feet by the wall height to get the total wall area (in square feet).

Once you know the total wall area, you can calculate the number of panels required. If you're using 4' x 8' fiberboard or plywood sheathing panels, divide the total wall area by 32. For example, if you need sheathing to cover 2,858 SF of wall area, divide the total area (2,858) by 32 to get 89.31 (or 90) wall sheathing panels.

You can also use Figure 7-5, in Chapter 7, to calculate the number of sheathing panels required.

Nails

Figure 8-4 is a factor table for estimating the quantity of nails you'll need for framing lumber and wall sheathing for an on-site wall system. This table includes an allowance for 15% waste. Here's how to use the table.

Framing lumber— Look at Figure 8-4 A. In the first column, enter the total board feet of framing lumber. Let's say our sample house requires 6,240 BF of framing lumber.

Multiply the total board feet (6,240) by the nail factor (0.02200). This gives us 137.28 lbs. of nails. Round up to 138. This allows for 22 lbs. of 16d common nails per MBF of framing lumber.

To estimate the cost of the nails, multiply the total number of pounds of nails by the cost per pound. If 16d nails cost $0.39 per pound, multiply the total number of pounds (138) by the cost per pound ($0.39) to get a total cost of $53.82.

Total board feet	x	Factor	=	Pounds nails
6,240	x	.02200	=	137.28 OR 138

Allow 22 pounds 16d common nails per 1,000 board feet.

A Studs, Plates and Headers

Total square feet	x	Factor	=	Pounds nails
2,240	x	.01000	=	22.40 OR 23

Allow 10 pounds 1½" roofing nails per 1,000 square feet.

B Fiberboard Sheathing.

Total square feet	x	Factor	=	Pounds nails
640	x	.00800	=	5.12 OR 6

Allow 8 pounds 6d threaded nails per 1,000 square feet.

C Plywood Sheathing

Factors for nails
Figure 8-4

Fiberboard sheathing— Look at Figure 8-4 B. In the first column, enter the total square feet of fiberboard sheathing. Let's say we need 2,240 SF of 4' x 8' x 1/2" fiberboard sheathing.

Multiply the total square feet (2,240) by the nail factor (0.01000). This gives us 22.4 lbs. of nails. Round up to 23. This allows for 10 lbs. of 1½" roofing nails per MSF of fiberboard sheathing.

To estimate the cost of the nails, multiply 23 lbs. by $0.56 (per lb.) to get a total cost of $12.88.

Plywood sheathing— Look at Figure 8-4 C. In the first column, enter the total square feet of plywood sheathing. In our example, we'll use 640 SF of 4' x 8' x 1/2" C-D plywood.

Multiply the total square feet (640) by the nail factor (0.00800). This gives us 5.12 lbs. of nails. Round up to 6. This allows for 8 lbs. of 6d threaded nails per MSF of plywood sheathing.

To estimate the cost of the nails, multiply 6 lbs. by $1.35 (per lb.) to get a total cost of $8.10.

Now add together the cost of nails for the framing lumber, fiberboard sheathing and plywood sheathing. The total cost of the nails for our sample wall system is $74.80.

Wall System Labor

The skill and productivity of your carpenters, the design of the house, and the weather, will all affect the cost of wall system labor. Be sure to allow for these variables in your labor estimate. Remember to use your cost records as your guide. Your own records are the most accurate source of labor costs for estimating any job.

Figures 8-5A and 8-5B are weekly time sheets for the Brown job. These time sheets show the manhours required to erect a preassembled wall system. Figures 8-6A and 8-6B are weekly time sheets for the Lawson job. These time sheets show the manhours required to erect a wall system that was assembled on the job site. Let's compare the labor for the two jobs.

Preassembled Wall System

The Brown job was a two-story factory-built house. The exterior walls were preassembled in panels not exceeding 16' in length for the first floor, and 14' for the second floor. The wall sheathing and windows (except the large picture window) were already in place. The door openings were framed in. The upper top plates were shipped loose.

The interior wall panels had all doors framed in. The first floor had 1,230 square feet of living area. The second floor had 1,040 square feet.

The daily log shows that on Friday, September 17, 40 manhours were required to erect the outside and inside wall sections on the first floor. On Monday, September 20, 40 manhours were required to place the upper top plates. That's a total of 80 manhours to erect the exterior and interior wall panels and place the upper top plates for the first floor.

On Thursday, September 23, the carpenters spent 20 manhours erecting the outside wall panels on the second floor. On Friday, September 24, 36 manhours were required to erect the remaining exterior and interior wall panels and upper top plates. The wall sections were then ready for the roof trusses. The second floor required a total of 56 manhours to erect the exterior and interior wall panels and place the upper top plates.

We've seen how many manhours were required to erect the preassembled wall system for the first and second floors of the Brown job. Now let's use this information to compute the manhour factors for a preassembled wall system. We'll use these manhour factors to estimate wall system labor for future jobs.

Weekly Time Sheet

For period ending 9-18-XX   *Brown* job

	Name	Exemptions	Days SEPTEMBER						Rate	Hours worked		Total earnings
			13 M	14 T	15 W	16 T	17 F	18 S		Reg.	Over-time	
1	C.A. Lester		X	8	5½	3	8	X		24½		
2	D.A. White		X	8	5½	3	8	X		24½		
3	A.L. King		X	8	5½	2½	8	X		24		
4	J.E. King		X	8	5½	2½	8	X		24		
5	W.R. Farlow		X	8	5½	2½	X	X		16		
6	J.R. Davis		X	8	5½	2½	8	X		24		
7												
8												
9												
10												
11												
12												
13												
14												
15												
16												
17												
18												
19												
20												

Daily Log

Monday XXXXX --- Factory-built house due tomorrow.

Tuesday Floor system on 1st floor (48 manhours)

Wednesday Finished subfloor on 1st floor (33 manhours) --- 1230 sq. ft.

Thursday Unloaded second trailer --- Rain (16 manhours)

Friday Set o/s and i/s partitions on 1st floor (1230 sq. ft. --- 40 manhours)

Saturday XXXX

Weekly time sheet — Brown job
Figure 8-5A

Weekly Time Sheet

For period ending __9-25-XX__   __Brown__ job

	Name	Exemptions	Days SEPTEMBER 20 M	21 T	22 W	23 T	24 F	25 S	Rate	Hours worked Reg.	Over-time	Total earnings
1	C.A. Lester		8	8	8	8	8	X		40		
2	D.L. White		8	8	8	8	8	X		40		
3	A.L. King		8	8	8	8	4	X		36		
4	W.R. Farlow		8	8	4	4	8	X		32		
5	J.R. Davis		8	8	8	8	8	X		40		
6	R.R. Lewis		X	8	4	4	8	X		24		
7												
8												
9												
10												
11												
12												
13												
14												
15												
16												
17												
18												
19												
20												

Daily Log

Monday   *Top plates on 1st floor partitions (40 Manhours)*

Tuesday   *Set beams and started floor joists on 2nd floor (48 Manhours)*

Wednesday   *Finished floor joists and bridging on 2nd floor (40 Manhours)*

Thursday   *Finished subfloor (1040 sq. ft.) (20 Manhours on 2nd floor) -- o/s partitions (20 Manhours)*

Friday   *Finished o/s and i/s partitions on 2nd floor (1040 sq. ft.) (36 Manhours) - Roof Trusses (8 Manhours)*

Saturday   *X X X*

Weekly time sheet — Brown job
Figure 8-5B

Weekly Time Sheet

For period ending **3-3-XX**   **LAWSON** job

	Name	Exemptions	Days Feb. / March						Rate	Hours worked		Total earnings	
			26 M	27 T	28 W	1 T	2 F	3 S		Reg.	Overtime		
1	R. R. Lewis		8	X	8	8	8	X		32			
2	C. A. Lester		2	X	8	X	X	X		10			
3	D. L. White		2	X	X	X	X	X		2			
4	R. C. Jones		8	X	8	8	8	X		32			
5	J. R. Davis		8	X	X	X	8	X		16			
6													
7													
8													
9													
10													
11													
12													
13													
14													
15													
16													
17													
18													
19													
20													

Daily Log

1682 SQ. FT.

Monday *Finished subfloor (14 manhours) --- o/s plates (14 manhours).*

Tuesday *Rain*

Wednesday *Changed basement stair opening (8 manhours) -- Laid out rooms (16 manhours)*

Thursday *Laid out windows and doors, cut studs for corner posts and tees (16 manhours)*

Friday *Assembled and emplaced o/s and i/s walls and partitions (24 manhours)*

Saturday *X X X X X*

Weekly time sheet — Lawson job
Figure 8-6A

Weekly Time Sheet

For period ending **3-10-XX**   **LAWSON** job

	Name	Exemptions	Days MARCH 5 M	6 T	7 W	8 T	9 F	10 S	Rate	Hours worked Reg.	Over-time	Total earnings
1	R. R. Lewis		6	8	7	8	8	X		37		
2	C. A. Lester		7	4	5½	8	8	X		32½		
3	R. C. Jones		6	8	7	8	8	X		37		
4	J. R. Davis		X	4	5½	8	8	X		25½		
5	D. L. White		X	4	5½	8	8	X		25½		
6												
7												
8												
9												
10												
11												
12												
13												
14												
15												
16												
17												
18												
19												
20												

Daily Log

Monday *Rain until 10:00 A.M. -- Framed in windows (19 manhours).*

Tuesday *Finished framing in window and doors -- Top plates (28 manhours)*

Wednesday *Installed wall sheathing (30½ manhours) 1682 sq.ft. on 1st floor.*

Thursday *Roof trusses (40 manhours)*

Friday *Finished roof trusses (10 manhours) -- Started roof sheathing (30 manhours)*

Saturday *X X X*

Weekly time sheet — Lawson job
Figure 8-6B

Manhour factors for preassembled wall system— Figure 8-7 A is a labor worksheet for the Brown job.

In the first column, enter the total manhours required to erect the preassembled wall system. The first floor required 80 hours. The second floor required 56 hours.

In the next column, enter the floor area (in square feet). The first floor had 1,230 SF of floor area. The second floor had 1,040 SF.

Now divide the total manhours (80 and 56) by the total floor areas (1,230 SF and 1,040 SF) to get the manhour factors. The manhour factor for the first floor is 0.06504. The factor for the second floor is 0.05385.

Notice that the manhour factor for the second floor is 17.2% less than the manhour factor for the first floor. Why? Because there are more partitions on the first floor, and the second-floor layout is simpler.

Use these manhour factors as a guide to future estimates. To find the number of manhours required for your new job, just multiply the manhour factor (for each floor) by the number of square feet of floor area.

Total manhours	÷	Floor area (square feet)	=	Manhour factor
1ST FLOOR 80	÷	1,230	=	.06504
2ND FLOOR 56	÷	1,040	=	.05385

A Preassembled wall system

Total manhours	÷	Floor area (square feet)	=	Manhour factor
147.5	÷	1,682	=	.08769

B On-site wall system

Wall system labor worksheet
Figure 8-7

On-site Wall System

Look back to Figures 8-6A and 8-6B. The weekly time sheets on the Lawson job show the manhours required to erect an on-site wall system. This job was a custom-built, one-story house. All of the framing lumber for the exterior and interior walls was cut and assembled on the job. The house had 1,682 SF of living area.

The daily log shows that on Monday, February 26, 14 manhours were required to place the sole plates for the outside walls. On Tuesday, it rained. No work was done. On Wednesday, carpenters spent 16 manhours laying out the rooms and installing the sole plates. On Thursday, March 1, another 16 manhours were required to lay out the windows and doors, cut the studs and assemble the corner post and ties. On Friday, March 2, workers assembled and placed the outside and inside wall panels. This required 24 manhours.

On Monday, March 5, it rained until 10:00 a.m. After that, the crew spent 19 manhours framing the windows. On Tuesday, 28 manhours were required to finish framing the windows and doors and install the top plates. On Wednesday, the crew spent 30½ manhours installing wall sheathing.

This work on the Lawson job required a total of 147½ manhours. Here's how to compute the manhour factor for an on-site wall system:

Manhour factor for on-site wall system— Look at Figure 8-7 B. In the first column, enter the total manhours. In our example, this comes to 147.5. In the next column, enter the total SF of floor area (1,682). Then divide the total manhours (147.5) by the total SF of floor area (1,682). The manhour factor for the on-site wall system is 0.08769.

You'll recall that the manhour factor for erecting the preassembled wall system on the first floor of the Brown job was 0.06504. The manhour factor for the Lawson job is 34.8% higher than the factor for the Brown job. The preassembled wall system required less labor.

Wall System Estimate (Sample)

We've seen how to use our cost records as a guide to easy, accurate wall system estimates. Now let's apply what we've learned. Let's calculate the materials and labor for the following problem.

We'll look at a one-story house with 2,270 SF of floor area. We'll compare the materials and labor for a preassembled wall system and the materials and labor for an on-site wall system.

Preassembled Wall System

First figure your material costs. Then compute your labor costs.

Materials— The materials for a preassembled wall system are the same as the materials for an on-site wall system, except that wall sheathing is usually fastened with staples instead of nails.

For our sample house with 2,270 SF of floor area, the wall system manufacturer looks at our blueprints and quotes us a price of $4,175.00. This includes:

1) Openings for windows and doors are to be cut out. (But no windows or doors are furnished.)

2) Sheathing is 4' x 8' x 1/2'' fiberboard, except corners. Corners are made of 4' x 8' x 1/2'' C-D plywood.

3) Delivery of the wall system to the job site.

Our total cost for materials for the preassembled wall system comes to $4,175.00 (plus sales tax, where required).

Labor— To estimate the labor required to erect the preassembled wall system, multiply the first-floor manhour factor from Figure 8-7 A (0.06504) by the floor area (2,270 SF) to get 147.64 manhours. Round up to 148. Multiply the total manhours (148) by the hourly wage rate ($25.00) and you have $3,700.00. This is your labor estimate for the preassembled wall system.

Our total cost for materials and labor for the preassembled wall system comes to $7,875.00. Now let's look at the on-site wall system.

On-site Wall System
Again, calculate your materials first and then your labor.

Materials— Remember that your on-site wall system materials will include:

1) Framing lumber— Be sure to include sole and top plates, studs, temporary braces, interior and exterior headers:

• Sole and top plates: In our sample house, there are 716 LF of exterior and interior walls. Multiply the total linear feet (716) by the number of plates (3) to get a total of 2,148 LF of lumber required.

We plan to use 2 x 4 x 12's for the sole and top plates. Now divide the total linear feet of lumber (2,148) by the lumber length (12) to get 179 pieces of lumber required to do the job.

To calculate the cost of the lumber, use Figure 8-1. Look down the first column and find the lumber size (2 x 4 x 12). In the next column, enter the number of pieces of lumber (179). Multiply 179 by the factor (0.0080) to get 1.432. Then multiply 1.432 by the price per MBF ($310.00) to get a total lumber cost of $443.92 for the sole and top plates.

• Studs: Allowing one stud for each linear foot of wall space, we'll need 716 studs for our sample house.

We'll use 2 x 4 x 8 lumber for the studs. To calculate the cost of the studs, use Figure 8-1. Multiply the number of pieces (716) by the factor

for 2 x 4 x 8 lumber (0.0053333) to get 3.81864. Then multiply 3.81864 by the price per MBF ($310.00) to get a total lumber cost of $1,183.78 for the studs.

Keep in mind that 2 x 4's are widely used on every job. When you order the lumber for studs, you'll probably want to order extra.

• Temporary braces: Allow 20 pieces of 2 x 4 x 12 lumber for braces. Use Figure 8-1 to calculate the cost. The total cost for temporary braces comes to $49.60.

• Interior and exterior headers. Our blueprints for our sample house indicate door widths and rough stud openings that will require the following headers:

Number of pieces	Lumber size
2	2 x 12 x 12
4	2 x 10 x 10
2	2 x 8 x 8
7	2 x 6 x 12
2	2 x 6 x 10
2	2 x 6 x 8
9	2 x 4 x 12
3	2 x 4 x 10
6	2 x 4 x 8

Using Figure 8-1 to calculate the cost of interior and exterior header lumber, we get a total lumber cost of $125.01.

To find the total cost of framing lumber for our sample house, add together the cost of sole and top plates ($443.92), studs ($1,183.78), temporary braces ($49.60), and interior and exterior headers ($125.01). This comes to a total cost of $1,802.31 for framing lumber.

2) Wall sheathing— Our blueprints tell us that we'll be using 4' x 8' x 1/2'' fiberboard and 4' x 8' x 1/2'' C-D plywood for wall sheathing. To figure the quantity of wall sheathing required, first multiply the total linear feet of exterior wall by the wall height to get a total wall area of 2,880 SF.

Now divide the total wall area (2,880 SF) by 32 (for 4' x 8' panels) to get a total of 90 panels required. (You can also use Figure 7-5, in Chapter 7, to calculate the number of sheathing panels required.) Of the 90 panels required, we'll need 70 fiberboard panels and 20 plywood panels. This comes to a total cost of $435.50.

3) Nails— Use Figure 8-4 to compute the quantity of nails required for our sample house.

Use Figure 8-4 A to calculate the quantity of nails required for framing lumber. In the first column, enter the total board feet of lumber required. In our sample house, this comes to 6,240 BF. This includes sole and top plates, studs, temporary braces, and interior and exterior headers.

Now multiply the total board feet (6,240) by the factor (0.02200) to get a total of 137.28 (or 138) lbs. of 16d common nails required for framing.

Use Figure 8-4B to compute the quantity of nails for fiberboard sheathing. We'll need 2,240 SF of fiberboard sheathing. Multiply the total square feet (2,240) by the factor (0.01000) to get a total of 22.40 (or 23) lbs. of 1½'' roofing nails required to do the job.

Figure 8-4 C is for plywood sheathing nails. We'll need 640 SF of plywood sheathing. Multiply the total square feet (640) by the factor (0.00800) to get a total of 5.12 (or 6) lbs. of 6d threaded nails required.

The total cost of nails for our sample wall system comes to $74.80. Now let's look at the cost for all materials for our sample on-site wall system.

Framing lumber comes to $1,802.31. Wall sheathing will cost $435.50. The cost of nails is $74.80. Add these together for a total materials cost of $2,312.61. Now let's look at labor for an on-site wall system.

Labor— To estimate labor for the on-site wall system, multiply the on-site manhour factor (0.08769) by the floor area (2,270 SF) to get 199.06 manhours. Round this down to 199. Multiply 199 by $25.00 to get $4,975.00. This is your labor estimate for the on-site wall system.

Our total cost for materials and labor for the on-site wall system comes to $7,287.61.

Now let's compare the two wall systems. Total cost for the preassembled wall system is $7,875.00. Total cost for the on-site wall system is $7,287.61. For our sample one-story house with a total floor area of 2,270 SF, it will be more profitable to use the on-site wall system.

A book, *Estimating Home Building Costs*, published by Craftsman Book Company, 6058 Corte del Cedro, Carlsbad, California 92008, explains in detail how to estimate all materials for a wall system in any house.

The wall system is only one phase of the project. But when these and other savings are combined, your total savings can be major. A few hundred dollars saved here and there on different parts of the job can make the difference between success and just surviving in the construction industry. Take advantage of savings like this to build your profits and reputation as a progressive and cost-conscious builder.

In the next chapter, we'll learn how to use our cost record system to increase our savings on roof system materials and labor.

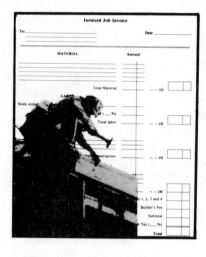

Chapter 9

Roof System Records

Roof trusses reduce framing costs. Use them if you can. You can put trusses up much faster than you can put up ceiling joists and rafters stick by stick. The labor savings can be major.

But it's not always possible to use roof trusses. The roof design and pitch may make conventional ceiling joists and rafters the only practical alternative.

In this chapter, we'll compare these two roof framing methods. We'll look at the materials and labor required for installing roof trusses, ceiling joists and rafters, sheathing and roof covering. We'll use tables and worksheets that will help us make quick, accurate roof system estimates.

Roof System Materials

Important components of the roof system include: framing (roof trusses or ceiling joists and rafters), sheathing, roof covering and nails.

Roof Trusses

Roof trusses are normally spaced 24" o.c. To calculate the number of roof trusses required, first find the length of the wall that's perpendicular to the truss. Multiply the wall length by 0.50, and add one truss for the starter.

For example, let's calculate the number of trusses required for a house with a 40' wall length. Multiply the wall length (40') by 0.50 to get 20 trusses. Add one truss for the starter. This house will require a total of 21 trusses.

If the roof line is broken into several sections, remember that you'll need to calculate trusses for each section of the roof.

Ceiling Joists and Rafters

Ceiling joists are normally spaced either 16" o.c. or 24" o.c. Use your normal estimating procedure to calculate the material required for ceiling joists.

If joists are spaced 16" o.c., multiply the total linear feet of wall by 0.75, and add one joist for the starter. If joist spacing is 24" o.c., multiply the total linear feet of wall by 0.50, and add one joist for the starter. The procedure is the same for both hip and gable roofs.

Estimating the material required for rafters is more complicated. On a sloped roof, the rafter length will always be longer than the horizontal distance covered by the rafter.

Some builders scale rafter lengths off an elevation view on the blueprints. But blueprints sometimes shrink, making the scaled distance inaccurate. And on some jobs, you won't have a good elevation view for the entire roof.

Let's look at the easy way to compute rafter lengths. We'll look at common rafters, hip and valley rafters, and shed roof rafters.

Common rafters— Figure 9-1 illustrates the terms you'll need to know to calculate the lengths of common rafters. These terms include: *span, run, rise, true rafter length* and *net rafter length.* Here's the meaning of each of these terms.

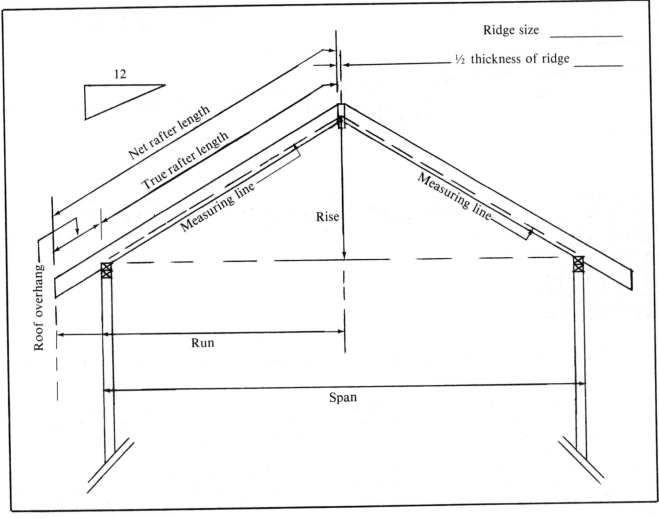

Common rafters
Figure 9-1

1) Span: The distance between supporting members of the roof.

2) Run: One-half the span. For rafters with an overhang, the run includes the distance from the wall to the tail cut of the rafter.

3) Rise: The vertical distance from the plate line to the top of the measuring line (sloping dotted line).

4) True rafter length: The distance between the face of the ridge and the outer face of the wall framing.

5) Net rafter length: The distance between the face of the ridge and the tail cut, or the total length of the rafter.

Now look at Figure 9-2. This is a factor table for cutting common rafters. You'll like the way it simplifies calculations. Here's how it works:

Suppose the roof pitch is 3½'' of rise per 12'' of run. And the span of the building is 26'0''. The run will be one-half the span, or 13'0'' (13.0').

Look down the first column to find the correct pitch (3½/12). In the next column, enter the run (13.0'). Multiply the run by the run factor (1.04167). Then subtract the ridge thickness factor (0.06250). This leaves 13.479' for the *true rafter length*.

To find the equivalent measurement in feet and inches, refer to Figure 9-3, a table of decimal equivalents of fractional parts of a foot. Here's how to use this table.

The first three columns show fractions of an inch. The first column shows 1/4'', 2/4'' and 3/4''. The second column shows eighths of an

Roof pitch	Run	x	Run factor	— (minus)	Factor for ½ thickness of ridge *	=	True rafter length **
1/12		x	1.00347	-	.06250	=	
1½/12		x	1.00778	-	.06250	=	
2/12		x	1.01379	-	.06250	=	
2½/12		x	1.02147	-	.06250	=	
3/12		x	1.03078	-	.06250	=	
3½/12	13.0'	x	1.04167	-	.06250	=	13.479' (13' 5¾")
4/12		x	1.05409	-	.06250	=	
5/12	12.573'	x	1.08333	-	.06250	=	13.558' (13' 6¹¹⁄₁₆")
6/12		x	1.11803	-	.06250	=	
7/12		x	1.15770	-	.06250	=	
8/12		x	1.20185	-	.06250	=	
9/12		x	1.25000	-	.06250	=	

* When ridge is 2" nominal framing (2 x 6's, 2 x 8's, etc.)
** True rafter length plus roof overhang equals net rafter length

Factors for common rafters
Figure 9-2

inch: 1/8", 2/8", 3/8", up through 7/8". The third column shows sixteenths of an inch: 1/16", 2/16", 3/16", 4/16", up through 15/16".

The remaining columns show whole inches, from 0" through 11". Let's try a sample calculation.

In our example above, the true rafter length comes to 13.479'. Now look at Figure 9-3. Find 0.479, and read up the column to the whole-inch number at the top (5"). Look again at 0.479 in the table. Now read across the column to the fractional inch number at the far left (3/4"). This table shows us that 0.479' is the same as 5¾". And our true rafter length of 13.479' is the equivalent of 13'5¾".

The *net rafter length* is the true rafter length plus the overhang. If the plans show the roof overhang in horizontal measurements, you compute the net rafter length by adding the run of the roof overhang to the run of the house. Then multiply the total by the run factor and subtract the ridge thickness factor. This gives you the net rafter length.

For example, say the roof pitch is 3½" for each 12" of run. The house run (to the exterior face of the wall) is 13'0", and the run of the roof overhang is 1'0". The total run is 14'0" (14.0'). Multiply 14.0' by the run factor (1.04167) to get 14.583'. Subtract the ridge thickness factor (0.06250) to get a net rafter length of 14.521'. To convert this to feet and inches, refer to the table in Figure 9-3. The net rafter length is 14'6¼".

Of course, not all spans are given in even feet. Usually the span will be given in feet, inches and fractional parts of an inch. For example, let's say a building has a roof pitch of 5 in 12 and a span of 25'1¾" (25.146'). The run for this span will be 12'6⅞" (12.573'). (Use the table of equivalents shown in Figure 9-3 to make your decimal conversions.)

To calculate the true rafter length for this example, multiply the run (12.573') by the run factor (1.08333) to get 13.621'. Subtract the ridge thickness factor (0.06250) to get a true rafter length of 13.558' (13'6¹¹⁄₁₆").

If the plans show the run of the roof overhang to be 2'5" (2.417'), compute the net rafter length by adding the run of the roof overhang (2'5") to the run of the house (12'6⅞"). This gives a total run of 14'11⅞" (14.990'). Multiply the total run (14.990')

4th	8th	16th	0"	1"	2"	3"	4"	5"	6"	7"	8"	9"	10"	11"
		0	.000	.083	.167	.250	.333	.417	.500	.583	.667	.750	.833	.917
		1	.005	.089	.172	.255	.339	.422	.505	.589	.672	.755	.839	.922
	1	2	.010	.094	.177	.260	.344	.427	.510	.594	.677	.760	.844	.927
		3	.016	.099	.182	.266	.349	.432	.516	.599	.682	.766	.849	.932
1	2	4	.021	.104	.188	.271	.354	.438	.521	.604	.688	.771	.854	.938
		5	.026	.109	.193	.276	.359	.443	.526	.609	.693	.776	.859	.943
	3	6	.031	.115	.198	.281	.365	.448	.531	.615	.698	.781	.865	.948
		7	.036	.118	.203	.286	.370	.453	.536	.620	.703	.786	.870	.953
2	4	8	.042	.125	.208	.292	.375	.458	.542	.625	.708	.792	.875	.958
		9	.047	.130	.213	.297	.380	.464	.547	.630	.714	.797	.880	.964
	5	10	.052	.135	.219	.302	.386	.469	.552	.635	.719	.802	.885	.969
		11	.057	.141	.224	.307	.391	.474	.557	.641	.724	.807	.891	.974
3	6	12	.063	.146	.229	.313	.396	.479	.563	.646	.729	.813	.896	.979
		13	.068	.151	.234	.318	.401	.484	.568	.651	.734	.818	.901	.984
	7	14	.073	.156	.240	.323	.406	.490	.573	.656	.740	.823	.906	.989
		15	.078	.161	.245	.328	.411	.495	.578	.661	.745	.828	.911	.995

Example: 8¼" = .688'

Decimal equivalents of fractional parts of a foot
Figure 9-3

by the run factor (1.08333) to get 16.239'. Then subtract the ridge thickness factor (0.06250) to get a net rafter length of 16.177' (16'2⅛").

Hip and valley rafters— Figure 9-4 is a factor table for computing the length of hip and valley rafters. It will help simplify your calculations. Here's how it works:

Let's say the roof has a pitch of 8 in 12 and a common rafter length of 14.0'. Look down the first column to find the correct pitch (8/12). In the next column, enter the run of the common rafter (14.0'). Multiply the run of the common rafter (14.0') by the factor (1.56347) to get a hip or valley rafter length of 21.889' (21'10¹¹⁄₁₆"). This is the length of the hip or valley rafter to the plate.

If there's an overhang, be sure to include it in your calculation. If the run of the common rafter overhang is 2'0", the total common rafter run will be 16'0" (14'0" plus the 2'0" overhang). Multiply the total run (16.0') by the factor for an 8/12 pitch (1.56347). This gives a net hip or valley rafter length of 25.016' or 25'0³⁄₁₆". (See Figure 9-3 for decimal conversion.)

Shed roof rafters— Figure 9-5 shows a shed roof rafter. Figure 9-6 is the factor table for estimating shed roof rafters. Here's how to use the table.

Let's say we have a porch with a roof pitch of 3½ in 12 and a porch run of 12'0" (12.0'). Look down the first column to find the correct pitch (3½/12). In the next column, enter the length of the porch run (12.0'). Multiply the run (12.0') by the factor (1.04167) to get 12.500'. Then subtract the nailer thickness factor (0.12500) to get a *true rafter length* of 12.375' (12'4½").

If the porch roof has an overhang run of 0'6" (0.5'), the total run of the roof would be 12'6" (12.5'). Use the total run of the roof to compute the *net rafter length*. Here's how:

Look down the first column to find the correct pitch (3½/12). In the next column, enter the length

Roof pitch	Run of common rafter	x	Factor	=	Length of hip or valley rafter (*)
3/12		x	1.43614	=	
3½/12		x	1.44398	=	
4/12		x	1.45297	=	
5/12		x	1.47432	=	
6/12		x	1.50000	=	
7/12		x	1.52980	=	
8/12	14.0'	x	1.56347	=	21.889' (21'10¹¹/₁₆")
9/12		x	1.60078	=	
10/12		x	1.64148	=	
11/12		x	1.68531	=	
12/12		x	1.73205	=	

* Add the rafter overhang to this length

Factors for hip and valley rafters
Figure 9-4

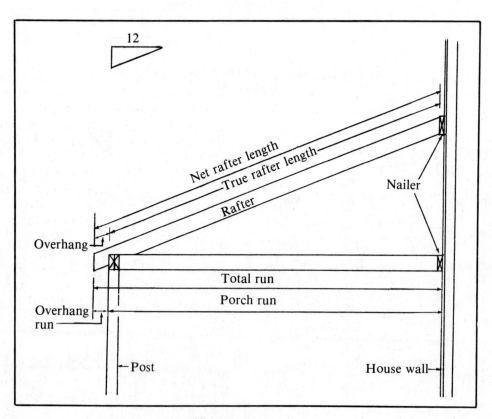

Shed roof rafters
Figure 9-5

Roof pitch	Run	x	Factor	— (minus)	Factor for full thickness of nailer *	=	True rafter length **
1/12		x	1.00347	-	.12500	=	
1½/12		x	1.00778	-	.12500	=	
2/12		x	1.01379	-	.12500	=	
2½/12		x	1.02147	-	.12500	=	
3/12		x	1.03078	-	.12500	=	
3½/12	12.0'	x	1.04167	-	.12500	=	12.375' (12'4½")
4/12		x	1.05409	-	.12500	=	
5/12		x	1.08333	-	.12500	=	
6/12		x	1.11803	-	.12500	=	
7/12		x	1.15770	-	.12500	=	
8/12		x	1.20185	-	.12500	=	
9/12		x	1.25000	-	.12500	=	

* When nailer is 2" nominal framing (2 x 6's, 2 x 8's, etc.)
** True rafter length plus overhang equals net rafter length

Factors for shed roof rafters
Figure 9-6

of the total run (12.5'). Multiply the length of the total run (12.5') by the factor (1.04167) to get 13.021'. Then subtract the nailer thickness factor (0.12500) to get a net rafter length of 12.896' (12'10¾").

Sheathing and Roof Covering
Once you know the rafter lengths, you can calculate the roof area. The net rafter length is the figure you'll use to compute the roof area. You'll use the roof area to estimate sheathing, roof covering and nails. Here's how to compute the roof area:

Suppose the net rafter length on a gable roof with no offsets is 14'8" (14.667'). Multiply the net rafter length (14.667') by the eave or ridge length (48.667') to get a roof area of 713.80 SF for one-half of the roof. Multiply this area (713.80 SF) by 2 to get a total roof area of 1,427.60 SF.

Now that you know the total roof area, you can estimate the sheathing, roof covering, and nails required for your roofing job.

Sheathing— To estimate the quantity of sheathing required, use the plywood panels factor table shown in Figure 7-5 in Chapter 7. In the first column, enter the total roof area (in square feet). Multiply the total area by the factor to get the number of sheathing panels required.

Roof covering— About 80% of all homes use asphalt or fiberglass shingles. Tile and wood shakes are popular in some areas, but fiberglass shingles are the choice of most builders.

Roofing materials are measured in *squares*. Each square equals 100 square feet of roof surface. To convert roof area to squares, divide the roof area (in square feet) by 100. Then add the extra shingles needed for starter strips, ridge and hip caps.

Figure 9-7 is a factor table for starter course shingles. Use Figure 9-7 A to compute the number of shingles required for the starter course. If you prefer to compute the number of squares required, use Figure 9-7 B. Both tables assume the use of 12" x 36" asphalt or fiberglass shingles. Here's how to use each table:

Suppose you're building a house with a total eave length of 128 linear feet. In the first column of Figure 9-7 A, enter the eave length. Multiply the eave length (128 LF) by the factor (0.33333) to get the number of shingles required for the starter course. In this example, it's 42.67, or 43 shingles.

Total eave length (linear feet)	x	Factor	=	Number of shingles
128	x	.33333	=	*42.67 or 43*

A   Shingles

Total eave length (linear feet)	x	Factor	=	Number of squares
128	x	.00417	=	*.53*

B   Squares

Factors for starter courses
Figure 9-7

Total length of ridge and hips (linear feet)	x	Factor	=	Number of shingles
52	x	.80000	=	*41.6 or 42*

A   Shingles

Total length of ridge and hips (linear feet)	x	Factor	=	Number of squares
52	x	.01000	=	*.52*

B   Squares

Factors for ridge and hip caps
Figure 9-8

Now look at Figure 9-7 B. In the first column, enter the eave length (128 LF). Multiply the eave length by the factor (0.00417) to get the number of squares of roofing you'll need for the starter course. In this example, it's 0.53 squares.

Figure 9-8 is a factor table for hip and ridge cap shingles. Use Figure 9-8 A to compute the number of shingles required for the hip and ridge caps. If you prefer to compute the number of squares required, use Figure 9-8 B. Both tables assume the use of 12'' x 36'' shingles with a 5'' exposure. Here's how to use each table:

Suppose a roof has 52 linear feet of ridge and hips. In the first column of Figure 9-8 A, enter the total length (52 LF). Multiply the length by the factor (0.80000) to get 41.6, or 42 shingles.

Now look at Figure 9-8 B. In the first column, enter the hip and ridge length (52 LF). Multiply the length by the factor (0.01000) to get the number of squares of roofing required for ridge and hip caps. In this example, it's 0.52 squares.

Add the number of squares (or shingles) needed for the ridge and hip caps to the squares (or shingles) needed for the starter course and the main roof. Then add about 3% for waste. Round the total up to the next whole number. This is the quantity of roof covering you'll need to order.

Nails
When calculating the quantity of nails required for your roofing job, refer back to Figure 7-6 in Chapter 7. And remember to include nails for both the main roof and for hip and ridge caps.

Main roof— Use Figure 9-9 to calculate the quantity of roofing nails required to shingle the main roof. Use Figure 9-9 A for 1'' nails and

Figure 9-9 B for 1¾'' nails. Both tables assume the use of 12'' x 36'' shingles. Here's how to use the tables:

For 1'' nails, enter the number of roofing squares in the first column of Figure 9-9 A. Let's say there are 29 squares. Multiply the number of squares (29) by the factor (1.5) to get 43.5 (44) lbs. of nails.

For 1¾'' nails, enter the number of roofing squares in the first column of Figure 9-9 B. Multiply the number of squares (29) by the factor (1.75) to get 50.75 (51) lbs. of nails.

Hip and ridge caps— Use Figure 9-10 to calculate the quantity of roofing nails required for ridge and hip caps. Use Figure 9-10 A for 1½'' nails and Figure 9-10 B for 1¾'' nails.

Number of squares	x	Factor	=	Number of pounds
29	x	1.5	=	*43.5 or 44*

A   1'' roofing nails

Number of squares	x	Factor	=	Number of pounds
29	x	1.75	=	*50.75 or 51*

B   1¾'' roofing nails

Factors for main-roof nails
Figure 9-9

Total length of ridge and hips (linear feet)	x	Factor	=	Number of pounds
67	x	.02667	=	*1.79 or 2*

A 1½" roofing nails

Total length of ridge and hips (linear feet)	x	Factor	=	Number of pounds
67	x	.03077	=	*2.06 or 2*

B 1¾" roofing nails

Factors for ridge and hip cap nails
Figure 9-10

For 1½" nails, enter the total ridge and hip length in the first column of Figure 9-10 A. Multiply the length (67 LF) by the factor (0.02667) to get 1.79 (2.0) lbs. of nails.

For 1¾" nails, enter the total ridge and hip length in the first column of Figure 9-10 B. Multiply the length (67 LF) by the factor (0.03077) to get 2.06 (2.0) lbs. of nails.

Roof pitch	Run of common rafter	x	Factor	=	Rise
1/12		x	.08333	=	
1½/12		x	.12500	=	
2/12		x	.16667	=	
2½/12		x	.20833	=	
3/12		x	.25000	=	
3½/12		x	.29167	=	
4/12		x	.33333	=	
5/12		x	.41667	=	
6/12	*14.5'*	x	.50000	=	*7.250' (7'3")*
7/12		x	.58333	=	
8/12		x	.66667	=	
9/12		x	.75000	=	

Factors for roof rise
Figure 9-11

Computing the Roof Rise

If the house has an attic, you'll need to know the rise of the roof. You'll use the rise to figure the available headroom in the attic. Figure 9-11 is a factor table for computing roof rise. Here's how to use it:

Let's say we have a house with a roof pitch of 6 in 12 and a common rafter run of 14'6" (14.5'). What's the rise of the roof?

Look down the first column to find the correct pitch (6/12). In the next column, enter the common rafter run (14.5'). Multiply the run (14.5') by the factor (0.50000) to get a roof rise of 7.250' (7'3").

This means that the vertical distance from the plate line to the top of the measuring line is 7'3". How much head room do we have in this attic? Plenty. The attic can be floored and used for storage space.

We've seen how to compute the materials required for our roofing job. Now let's take a look at roof system labor.

Roof System Labor

As with every estimate, use your cost records from previous jobs as your guide to accurate roof system labor estimates. There are many ways to compile the manhour factors shown in your cost records. Whether you keep your records in units of square feet or square yards isn't important. *But do keep a record of the manhours required to do every part of every job.*

Roof system labor includes the manhours required for: framing (roof trusses or ceiling joists and rafters), sheathing, and roof covering. Let's look at the labor required for both the truss roof system and the ceiling joist and rafter system.

Truss Roof System

Figures 9-12A, 9-12B and 9-12C are weekly time sheets for the Brown job. This is the same two-story house we've estimated in preceding chapters. The roof covers an area of 2,317 SF over the living quarters and porches. Roof trusses were used throughout, except for a small area over a kitchen-nook, where ceiling joists and rafters were used.

Our daily log for Friday, September 24, shows that 8 manhours were required to place the roof trusses over the kitchen and living room area. On Monday, September 27, carpenters spent 18 manhours installing a supporting beam, ceiling joists and rafters over the kitchen-nook. On Tuesday, September 28, it rained. No work was done.

Weekly Time Sheet

For period ending 9-25-xx   BROWN   job

#	Name	Exemptions	Days SEPTEMBER 20 M	21 T	22 W	23 T	24 F	25 S	Rate	Hours worked Reg.	Over-time	Total earnings
1	C. A. LESTER		8	8	8	8	8	X		40		
2	D. L. WHITE		8	8	8	8	8	X		40		
3	A. L. KING		8	8	8	8	4	X		36		
4	W. R. FARLOW		8	8	4	4	8	X		32		
5	J. R. DAVIS		8	8	8	8	8	X		40		
6	R. R. LEWIS		X	8	4	4	8	X		24		
7												
8												
9												
10												
11												
12												
13												
14												
15												
16												
17												
18												
19												
20												

Daily Log

Monday TOP PLATES ON 1ST FLOOR PARTITIONS (40 MANHOURS)

Tuesday SET BEAMS AND STARTED FLOOR JOISTS ON 2ND FLOOR (48 MANHOURS)

Wednesday FINISHED FLOOR JOISTS AND BRIDGING ON 2ND FLOOR (40 MANHOURS)

Thursday FINISHED SUBFLOOR (20 MANHOURS ON 2ND FLOOR) — O/S PARTITIONS (20 MANHOURS) 1,040 S.F.

Friday FINISHED O/S AND I/S PARTITIONS ON 2ND FLOOR (36 MANHOURS) — AND LIVING AREA (8 MANHOURS) ROOF TRUSSES OVER KIT. (624 SF)

Saturday X X X X X

Weekly time sheet
Figure 9-12A

Weekly Time Sheet

For period ending 10-2-xx   BROWN ___job

	Name	Exemptions	Days SEPT. 27 M	28 T	29 W	30 T	OCT. 1 F	2 S	Rate	Hours worked Reg.	Over-time	Total earnings
1	R. R. LEWIS		X	X	8	2	X	X		10		
2	C. A. LESTER		3	X	8	2	X	X		13		
3	D. L. WHITE		3	X	8	2	X	X		13		
4	A. L. KING		3	X	8	2	X	X		13		
5	J. E. KING		3	X	8	2	X	X		13		
6	W. R. FARLOW		3	X	8	2	X	X		13		
7	J. R. DAVIS		3	X	8	2	X	X		13		
8												
9												
10												
11												
12												
13												
14												
15												
16												
17												
18												
19												
20												

Daily Log

53 SF
(18 MANHOURS)

Monday BEAM, CEILING JOISTS AND RAFTERS OVER KITCHEN-NOOK

Tuesday RAIN

Wednesday ROOF TRUSSES ON 2ND FLOOR -- STARTED ROOF SHEATHING (56 MANHOURS) 1040 SF

Thursday STARTED PORCH TRUSSES -- RAIN (14 MANHOURS)

Friday RAIN

Saturday X X X X X

Weekly time sheet
Figure 9-12B

Weekly Time Sheet

For period ending 10-9-XX   BROWN   job

	Name	Exemptions	Days OCTOBER						Rate	Hours worked		Total earnings
			4 M	5 T	6 W	7 T	8 F	9 S		Reg.	Over-time	
1	R. R. LEWIS		8	8	X	X	X	X		16		
2	C. A. LESTER		4	8	8	8	8	X		36		
3	D. L. WHITE		8	8	8	8	8	X		40		
4	A. L. KING		8	8	X	8	X	X		24		
5	J. E. KING		8	8	X	8	X	X		24		
6	W. R. FARLOW		8	8	8	8	4	X		36		
7	J. R. DAVIS		8	8	8	8	8	X		40		
8												
9												
10												
11												
12												
13												
14												
15												
16												
17												
18												
19												
20												

Daily Log

Monday PORCH TRUSSES IN BACK -- STARTED FRONT PORCH TRUSSES (52 MANHOURS)
ROOFS OVER FRONT AND BACK PORCHES
Tuesday FINISHED FRONT PORCH TRUSSES -- ROOF SHEATHING (56 MANHOURS)
Wednesday FINISHED ROOF SHEATHING AND FELT ON HOUSE (32 MANHOURS)
Thursday SET WINDOWS AND FRONT DOOR — RAIN (48 MANHOURS)
Friday SET 3 DOORS IN BACK -- STAIRS TO 2ND FLOOR (28 MANHOURS)
Saturday X X X X

Weekly time sheet
Figure 9-12C

Total Manhours	÷	Floor area (square feet)	=	Manhour Factor
236	÷	2,317	=	.10186

A Trusses

Total Manhours	÷	Floor area (square feet)	=	Manhour Factor
592	÷	3,008	=	.19681

B Ceiling joists and rafters

Roof system labor worksheet
Figure 9-13

On Wednesday, September 29, the crew placed the roof trusses over the second floor and started installing the plywood roof sheathing. A total of 56 manhours were recorded. On Thursday, September 30, 14 manhours were required to set the roof trusses on the porches. On Friday, October 1, it rained again, and no work was done.

On Monday, October 4, carpenters spent 52 manhours finishing up the roof trusses on the back porch and starting the trusses for the front porch. On Tuesday, October 5, they finished the front porch and began on the sheathing. A total of 56 manhours were recorded. On Wednesday, October 6, they spent 32 manhours finishing the sheathing and applying felt.

A total of 236 manhours were required to frame the roof as described here. The floor area covered was 2,317 square feet. Now let's use this information to compute a manhour factor for installing trusses.

Manhour factor for truss roof system— Figure 9-13 is a roof system labor worksheet. Use the table in Figure 9-13 A to compute the manhour factor for roof trusses. The table shown in Figure 9-13 B is for computing the manhour factor for conventional ceiling joists and rafters.

Here's how to compute the manhour factor for the roof trusses on the Brown job. Enter the total manhours (236) in the first column of Figure 9-13 A. In the next column, enter the total square feet of floor and porch area (2,317). Now divide the total manhours (236) by the total floor and porch area (2,317) to get a manhour factor of 0.10186.

To estimate the manhours required for a truss roofing system similar to the Brown job, just multiply the manhour factor (0.10186) by the total

floor area. This gives you the total manhours required to do the job.

To estimate the cost of labor for the job, multiply the total manhours by the hourly rate of pay. This will be your labor estimate for the truss framing portion of the job.

Ceiling Joist and Rafter System
Figures 9-14A through 9-14E are weekly time sheets showing the manhours required for the roof system on the Green job. This house was a large two-story dwelling with an attached double garage. The roof was designed with many offsets. This made it impractical to use roof trusses. Ceiling joists and rafters had to be individually cut and placed.

According to the daily log for Friday, July 1, 32 manhours were required to place the ceiling joists on the second floor. Another 32 hours were needed to place the floor joists over the garage.

Monday, July 4, was a holiday. No work was done. On Tuesday, July 5, workers spent 32 manhours placing floor joists and bridging over the garage. On Wednesday, July 6, 28 manhours were required to set the gables and prepare to place the rafters. On Thursday, 48 manhours were required to place the rafters. Another 44 manhours were spent placing rafters on Friday, July 8.

On Monday and Tuesday of the following week, another 72 manhours were required to place rafters. On Wednesday, July 13, 32 manhours were required to finish placing the rafters. Workers spent 80 manhours applying plywood roof sheathing on Thursday and Friday of that week.

Another 96 manhours were required to complete the application of the roof sheathing. Roof sheathing was finished on Wednesday, July 20. On Thursday and Friday, workers spent 56 manhours installing roofing felt.

On Monday, July 25, 16 manhours were spent installing roofing felt. On Tuesday, workers finished installing the roofing felt and built a decorative roof overhang. This work required 24 manhours.

The total floor area covered on the Green job was 3,008 square feet. A total of 592 manhours were required to construct this roof system using ceiling joists and rafters. Let's use this information to compute a manhour factor for roof systems using ceiling joists and rafters.

Manhour factor for ceiling joist and rafter system— Look back to Figure 9-13 B. In the first column, enter the total manhours (592). In the next column, enter the total square feet of floor area (3,008). Then divide the total manhours (592) by the total floor area (3,008 SF) to get a manhour factor of 0.19681.

Weekly Time Sheet

For period ending ___7-2-XX___   ___GREEN___ job

	Name	Exemptions	Days JUNE 27 M	28 T	29 W	30 T	JULY 1 F	2 S	Rate	Hours worked Reg.	Over-time	Total earnings
1	D.L. WHITE		8	8	8	8	8	X		40		
2	C.A. LESTER		8	8	8	8	8	X		40		
3	A.L. KING		8	8	8	8	8	X		40		
4	J.E. KING		8	8	8	8	8	X		40		
5	D.L. WEST		8	8	8	8	8	X		40		
6	L.H. KIDD		8	8	8	8	8	X		40		
7	A.E. KIDD		8	8	8	8	8	X		40		
8	R.C. JONES		8	8	8	8	8	X		40		
9												
10												
11												
12												
13												
14												
15												
16												
17												
18												
19												
20												

Daily Log

Monday SET BALANCE OF PARTITONS ON 1ST FLOOR -- STRAIGHTENED WALLS. (40 MANHOURS) TOP PLATES -- STARTED FLOOR JOISTS ON 2ND FLOOR (24 MANHOURS)

Tuesday FLOOR JOISTS ON 2ND FLOOR -- BRIDGING (64 MANHOURS)

Wednesday CIRCLE STAIRS (32 MANHOURS) -- SUBFLOOR ON 2ND FLOOR (32 MANHOURS)

Thursday SET PARTITIONS ON 2ND FLOOR (64 MANHOURS)

Friday CEILING JOISTS ON 2ND FLOOR (32 MANHOURS) STARTED FLOOR JOISTS OVER GARAGE (32 MANHOURS)

Saturday ___X X X X X___

Weekly time sheet
Figure 9-14A

Weekly Time Sheet

For period ending ___7-9-XX___   ___GREEN___ job

	Name	Exemptions	Days JULY						Rate	Hours worked		Total earnings
			4 M	5 T	6 W	7 T	8 F	9 S		Reg.	Over-time	
1	C. A. LESTER		X	8	4	8	4	X		24		
2	4. L. KING		X	8	8	8	8	X		32		
3	J. E. KING		X	8	8	8	8	X		32		
4	D. L. WEST		X	8	8	8	8	X		32		
5	L. H. KIDD		X	X	X	8	8	X		16		
6	R. C. JONES		X	X	X	8	8	X		16		
7												
8												
9												
10												
11												
12												
13												
14												
15												
16												
17												
18												
19												
20												

Daily Log

Monday ___JULY 4TH HOLIDAY___
Tuesday ___FINISHED FLOOR JOISTS AND BRIDGING OVER GARAGE (32 MAN-HOURS)___
Wednesday ___SET GABLES -- PREPARATION WORK FOR RAFTERS (28 MAN-HOURS)___
Thursday ___RAFTERS (48 MAN-HOURS)___
Friday ___RAFTERS (44 MAN-HOURS)___
Saturday ___X X X X X___

Weekly time sheet
Figure 9-14B

Weekly Time Sheet

For period ending 7-16-XX   GREEN job

	Name	Exemptions	Days JULY 11 M	12 T	13 W	14 T	15 F	16 S	Rate	Hours worked Reg.	Over-time	Total earnings
1	A. L. KING		8	8	8	8	8	X		40		
2	J. E. KING		8	8	8	8	8	X		40		
3	D. L. WEST		8	8	X	8	8	X		32		
4	L. H. KIDD		8	8	8	8	8	X		40		
5	R. R. LEWIS		X	8	8	8	8	X		32		
6												
7												
8												
9												
10												
11												
12												
13												
14												
15												
16												
17												
18												
19												
20												

Daily Log

Monday RAFTERS (32 MAN-HOURS)

Tuesday RAFTERS (40 MAN-HOURS)

Wednesday FINISHED RAFTERS (32 MAN-HOURS)

Thursday STARTED PLYWOOD ROOF SHEATHING (40 MAN-HOURS)

Friday PLYWOOD ROOF SHEATHING; FINISHED FRONT (40 MAN-HOURS)

Saturday XXXXX

Weekly time sheet
Figure 9-14C

Weekly Time Sheet

For period ending _7-23-XX_   _GREEN_ job

	Name	Exemptions	Days JULY						Rate	Hours worked		Total earnings
			18	19	20	21	22	23		Reg.	Over-time	
			M	T	W	T	F	S				
1	B.R. LEWIS		8	8	8	8	X	X		32		
2	C.4. LESTER		8	8	8	8	8	X		40		
3	D.L. WEST		8	8	8	8	8	X		40		
4	L.H. KIDD		8	8	8	8	8	X		40		
5												
6												
7												
8												
9												
10												
11												
12												
13												
14												
15												
16												
17												
18												
19												
20												

Daily Log

Monday _PLYWOOD ROOF SHEATHING (32 MANHOURS)_
Tuesday _PLYWOOD ROOF SHEATHING (32 MANHOURS)_
Wednesday _FINISHED PLYWOOD ROOF SHEATHING (32 MAN HOURS)_
Thursday _STARTED 15 LB FELT ON ROOF (32 MANHOURS)_
Friday _ROOF FELT (24 MANHOURS)_
Saturday _XXXXX_

Weekly time sheet
Figure 9-14D

Weekly Time Sheet

For period ending __7 - 30 - XX__   __GREEN__ job

	Name	Exemptions	Days JULY						Rate	Hours worked		Total earnings
			25 M	26 T	27 W	28 T	29 F	30 S		Reg.	Over-time	
1	C. A. LESTER		8	8	8	8	8	X	40			
2	D. L. WEST		8	8	X	X	8	X	24			
3	L. H. KIDD		X	8	8	8	8	X	32			
4												
5												
6												
7												
8												
9												
10												
11												
12												
13												
14												
15												
16												
17												
18												
19												
20												

Daily Log

Monday __ROOF FELT (16 MAN-HOURS)__
Tuesday __CONSTRUCTED DECORATIVE ROOF OVER-HANG --- FINISHED ROOF FELT__ (TOTAL FLOOR AREA FOR 2ND FLOOR AND GARAGE = 3008 SQ.FT. (24 MAN HOURS))
Wednesday __O/S TRIM (16 MAN-HOURS)__
Thursday __O/S TRIM (16 MAN-HOURS)__
Friday __O/S TRIM (24 MAN-HOURS)__
Saturday __XXXXX__

Weekly time sheet
Figure 9-14E

To estimate the manhours required for a ceiling joist and rafter system similar to the Green job, just multiply the manhour factor (0.19681) by the total floor area. This gives you the total manhours required to do the job.

To estimate the cost of labor for the job, multiply the total manhours by the hourly rate of pay.

Now let's compare the manhour factors for the two different roof systems. The manhour factor for the Green job using ceiling joists and rafters is 0.19681. The manhour factor for the Brown job, using roof trusses, is 0.10186. The manhour factor for the Green job is 93.2% higher than the manhour factor for the Brown job — almost double! If your own crew is installing the roof system, use trusses whenever possible.

Roofing can usually be installed much cheaper by a roofing subcontractor. An experienced roofer can install more than one square of asphalt or fiberglass shingles per hour on a gable roof if the pitch isn't over 6 in 12. If your own carpenters aren't experienced roofers, it may take them up to 6 manhours per square to do the same job.

If you don't have a skilled roofer on the payroll, get bids from several roofing subs. Using roofing subcontractors also releases your carpenters to do the work they do best.

In this chapter, we've looked at the easy way to make accurate roof system estimates. Next we'll learn how to make accurate insulation and wallboard estimates.

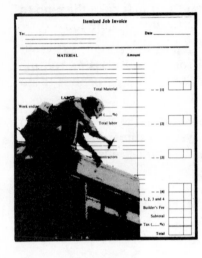

Chapter 10

Insulation and Wallboard Records

High energy costs make insulation important to every homeowner and every builder. All building materials have some insulating value. But there are materials available that are designed just for insulation. They provide an effective barrier against heat loss and heat gain. Install these materials in the homes you build. Quality insulation work will pay large dividends during the entire life of the home.

Before 1973, most houses had full insulation only in the ceilings. Little or no insulation was used in walls or floors. Storm sashes and storm doors were used only regionally. Caulking and sealant around wall openings was haphazard or nonexistent.

Attempts have been made to reduce energy consumption in older homes. For example, one homeowner added four inches of insulation in the attic, filled the walls with blown insulation and put batts under the floor. Storm sashes and storm doors were installed.

After the job was complete, the owner kept a record of energy required for heating for the next ten years. The result: Fuel consumption was reduced by an average of 46.19%. And fuel savings weren't the only benefit. The house was more comfortable and draft-free in cold weather. These benefits will continue as long as the house stands.

In this chapter, we'll look at insulation materials and labor. We'll also cover wallboard materials and labor. We'll begin with a look at R-values. Then we'll use our cost records as a guide to easy, accurate insulation and wallboard estimates.

R-values

The ability of an insulating material to resist heat transfer is indicated by its *R-value*. The higher the R-value, the more effective the insulation. Figure 10-1 shows how a "blanket of insulation" should be installed around a house for maximum energy conservation.

To determine the correct R-value, look at the climate of the area where the house is built. In warmer climates, use R-19 in ceilings and R-11 in walls and floors. In colder regions, use R-38 in ceilings, R-19 in walls and R-22 under floors.

Insulation Materials

Let's look at the insulation required for ceilings, walls and floors.

Ceilings

Ceiling insulation can be either batts, blankets, loose fill (poured-in) or a combination of these. Figure 10-2 shows the different types of insulation and how thick each must be to get the desired R-value.

Here's a sample ceiling heat-loss comparison:

1) 8-inch insulation (R-25.3) has 21% less heat loss than 6'' insulation (R-19.0).

2) 10-inch insulation (R-31.7) has 35% less heat loss than 6'' insulation (R-19.0).

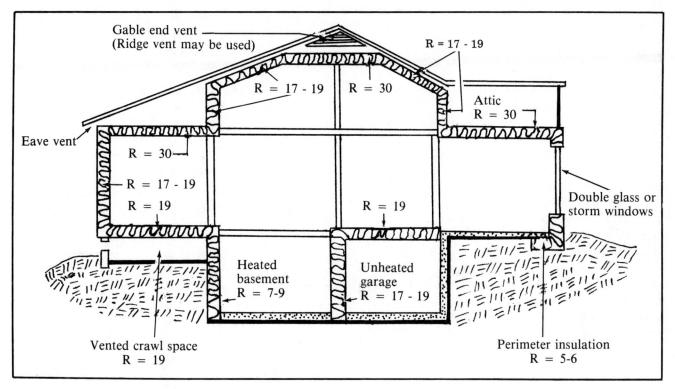

"Blanket of insulation"
Figure 10-1

3) 12-inch insulation (R-38) has 46% less heat loss than 6" insulation (R-19.0).

To calculate the material required for ceiling insulation, you'll need to know the total floor area. Your blueprints will tell you the total floor area. Divide the floor area by the number of square feet of insulation in each roll. This will tell you the number of rolls of insulation you'll need to order.

Here's a sample calculation. Suppose you're insulating a house with a total floor area of 1,040 SF and trusses spaced 24" o.c. (Insulation should be no wider than 23" for trusses or joists spaced 24" o.c.) If you want a combined R-value of 30, you can use one layer of R-19 insulation plus one layer of R-11 insulation.

R-19 insulation (6" x 23") normally comes 75 SF to a roll. Divide your total floor area (1,040 SF) by

R-value	Batts or blankets		Loose fill (poured-in)		
	Glass fiber	Rock wool	Glass fiber	Rock wool	Cellulosic fiber
R-11	3½" - 4"	3"	5"	4"	3"
R-19	6" - 6½"	5¼"	8" - 9"	6" - 7"	5"
R-22	6½"	6"	10"	7" - 8"	6"
R-30	9½" - 10½"*	9"*	13" - 14"	10" - 11"	8"
R-38	12" - 13"*	10½"*	17" - 18"	13" - 14"	10" - 11"

* Two batts or blankets required

R-value and insulation thickness
Figure 10-2

the number of square feet per roll (75) to get a total of 13.87 (or 14) rolls of R-19 insulation required.

If R-11 insulation (3½'' x 23'') comes 135.12 SF to a roll, divide the total floor area (1,040 SF) by 135.12. You'll need to order 7.70 (or 8) rolls of R-11 insulation for the job.

Walls

Exterior walls (excluding windows and doors) account for 13.2% of the total heat loss in a house. When R-11 or R-13 batt or blanket insulation is added to a wall, the result will be an R-17 to R-19 wall. A brick veneer wall with R-11 insulation between the studs will lose 69% less heat than the same wall with no insulation between the studs.

To compute the material required for wall insulation, you'll need to know the total area of all exterior and interior walls where insulation is required. To find the wall area, multiply the wall perimeter by the wall height. This gives you the gross area. Deduct any door or window openings that are 50 SF (or larger). This gives you the net area. Divide the net area by the amount of insulation in each bag or roll.

Here's an example. Let's insulate 1,968 SF of wall with studs spaced 16'' o.c. We'll use R-11 insulation (3½'' x 15''), which comes 88 SF to a roll. Divide the total wall area (1,968 SF) by the number of square feet per roll (88) to get a total of 22.36 (or 23) rolls of insulation required to do the job.

Floors

Floors over heated rooms or basements need no insulation. (See Figure 10-1.) If a floor is over an unheated garage, unheated basement or vented crawl space, use about R-19 insulation (depending on the climate).

To calculate the material required for floor insulation, you'll need to know the total area of the space where insulation is required. Remember: no insulation is required in flooring over heated rooms or basements. Divide the area by the number of square feet of insulation in each roll. This gives you the number of rolls you'll need to order.

Let's say you want to insulate a floor space of 1,040 SF. Joists are spaced 16'' o.c. Specs call for R-11 insulation (3½'' x 15''), which comes packed 88 SF per roll. Divide the floor area (1,040 SF) by the number of square feet per roll (88) to get 11.82 (or 12) rolls of R-11 insulation required.

The effectiveness of insulation depends also on the quality of the installation. Compressing batts or blankets increases the density and decreases the R-value. Blown insulation with too much air in the mixture results in a "fluffy" application. In time, the insulation settles and no longer has the thickness needed to provide the intended R-value.

Wallboard Materials

Gypsum drywall is the most popular finish material for walls and ceilings in residential construction. It's inexpensive, goes up quickly and requires little drying time. The walls provide a smooth, durable surface that can be finished with paint or fabric.

Begin your wallboard estimate by calculating the area of the walls and ceiling in each room, closet, hall and stairway.

Figure 10-3 shows the total square feet of walls and ceiling for rooms of different sizes. This table assumes a ceiling height of 8'0''. If a room is 14'0'' x 16'0'', for example, it will have 704 SF of wall and ceiling area. A hall 3'0'' x 18'0'' has 390 SF of wall and ceiling area.

Here's a tip. Estimate all walls as if they had no door or window openings. This will give you a small allowance for waste. And the piece of board cut out for a window or door probably can't be used elsewhere.

When the room is larger than the dimensions given in Figure 10-3, or when the ceiling is not 8' high, use Figure 10-4 to calculate the wall and ceiling area. Here's an example.

If you have a garage that is 22'0'' x 24'0'' and has a ceiling height of 10'0'', what is the total wall and ceiling area? Figure 10-4 shows that there are 1,448 SF of walls and ceiling in this garage.

If you want to convert the total square feet into square yards, simply divide by 9. In our example, 1,448 square feet divided by 9 equals 160.89 or 161 square yards.

Wallboard Panels

Once you know the total wall and ceiling area, you can figure the number of sheets of wallboard to order. Use Figure 10-5 A to compute the number of 4' x 8' panels required. Use Figure 10-5 B to compute the number of 4' x 12' panels required. Here's how to use these two factor tables.

In the first column, enter the total square feet of wall and ceiling area. In this example, there are 10,016 SF of surface area to be covered. Multiply the total surface area by the factor to get the number of panels required. Figure 10-5 A shows that if you use 4' x 8' panels, you'll need 313 panels to complete the job. If you're using 4' x 12' panels, Figure 10-5 B shows that you'll need only 209 panels.

Length of room

	2'	3'	4'	5'	6'	7'	8'	9'	10'	11'	12'	13'	14'	15'	16'	17'	18'	19'	20'
2'	68	86	104	122	140	158	176	194	212	230	248	266	284	302	320	338	356	374	392
3'	86	105	124	143	162	181	200	219	238	257	276	295	314	333	352	371	390	409	428
4'	104	124	144	164	184	204	224	244	264	284	304	324	344	364	384	404	424	444	464
5'	122	143	164	185	206	227	248	269	290	311	332	353	374	395	416	437	458	479	500
6'	140	162	184	206	228	250	272	294	316	338	360	382	404	426	448	470	492	514	536
7'	158	181	204	227	250	273	296	319	342	365	388	411	434	457	480	503	526	549	572
8'	176	200	224	248	272	296	320	344	368	392	416	440	464	488	512	536	560	584	608
9'	194	219	244	269	294	319	344	369	394	419	444	469	494	519	544	569	594	619	644
10'	212	238	264	290	316	342	368	394	420	446	472	498	524	550	576	602	628	654	680
11'	230	257	284	311	338	365	392	419	446	473	500	527	554	581	608	635	662	689	716
12'	248	276	304	332	360	388	416	444	472	500	528	556	584	612	640	668	696	724	752
13'	266	295	324	353	382	411	440	469	498	527	556	585	614	643	672	701	730	759	788
14'	284	314	344	374	404	434	464	494	524	554	584	614	644	674	704	734	764	794	824
15'	302	333	364	395	426	457	488	519	550	581	612	643	674	705	736	767	798	829	860
16'	320	352	384	416	448	480	512	544	576	608	640	672	704	736	768	800	832	864	896
17'	338	371	404	437	470	503	536	569	602	635	668	701	734	767	800	833	866	899	936
18'	356	390	424	458	492	526	560	594	628	662	696	730	764	798	832	866	900	934	968
19'	374	409	444	479	514	549	584	619	654	689	724	759	794	829	864	899	934	969	1004
20'	392	428	464	500	536	572	608	644	680	716	752	788	824	860	896	932	968	1004	1040

(Width of room — row labels down the left)

Total area (square feet) of walls and ceiling
(rooms with 8'0" ceiling height)
Figure 10-3

Room: *GARAGE*

Size: *24'0"* WIDTH X *22'0"*

Ceiling height: *10'0"*

Perimeter of walls: *24'* width + *22'* length x 2 = *92* linear feet

Walls: *92* LF (perimeter) x *10'* (ceiling height) .... *920* SF

Ceiling: *24'* width x *22'* length .... *528* SF

Total square feet of walls and ceiling .... *1448* SF*

1448 square feet divided by 9 = *160.89* square yards

*If openings are to be deducted, deduct them from this total.

Total area (square feet) of walls and ceiling
(rooms with ceiling height other than 8'0")
Figure 10-4

Total surface area (square feet)	x	Factor	=	Number of panels
10,016	x	.03125	=	*313*

A 4' x 8' wallboard

Total surface area (square feet)	x	Factor	=	Number of panels
10,016	x	.02083	=	*208.63 or 209*

B 4' x 12' wallboard

Factors for wallboard panels
Figure 10-5

Wallboard size	Approximate weight (each panel)	Bending Radii Length	Width
¼" x 4' x 8'	35 lbs.	5'	15'
⅜" x 4' x 8'	50 lbs.	7½'	25'
⅜" x 4' x 12'	75 lbs.	7½'	25'
½" x 4' x 8'	67 lbs.	10'*	---
½" x 4' x 12'	100 lbs.	10'*	---
⅝" x 4' x 8'	90 lbs.	---	---
⅝" x 4' x 12'	135 lbs.	---	---

* Bending two ¼" pieces successively permits radii shown for ¼".

Gypsum board weight and bending radii
Figure 10-6

When the wallboard is delivered to the job site, be aware of how much weight you're storing in each room. Figure 10-6 shows gypsum board weight and bending radii. Use this table to determine how much wallboard weight you're bringing onto the job site. Here's how the table works:

Using our example in Figure 10-5 B, we'll need to store 209 pieces of 1/2" x 4' x 12' gypsum wallboard. According to the chart in Figure 10-6, each piece of 1/2" x 4' x 12' gypsum wallboard weighs approximately 100 lbs. This makes a total weight of 20,900 lbs. of wallboard to be delivered and stored at the job site. Divide 20,900 lbs. by 2,000 to get 10.45 tons of wallboard.

When the wallboard arrives, divide it up and store it throughout the house. Don't put that 10 tons across two or three joists in a single room! And be sure to stack it so that the length of the wallboard is perpendicular to the floor joists. This distributes the weight over a greater number of joists.

Fasteners
Common wallboard fasteners are nails and adhesive. Use the nails together with the adhesive. Figure 10-7 A is a factor table for 1⅜" annular ring nails. Figure 10-7 B is a factor table for adhesive. Here's how to use these two tables.

In the first column, enter the total square feet of wallboard surface area. In our example, this is 10,016 SF. Then multiply the total area by the factor to get the number of pounds of nails and the number of tubes of adhesive required.

Figure 10-7 A shows that we'll need 50 lbs. of an-

nular ring nails. Figure 10-7 B shows that 20 tubes of adhesive are required.

Finishing Materials
Use Figure 10-8 to compute the quantity of finishing materials needed. Figure 10-8 A is a factor table for rolls of tape (based on 250' per roll). Figure 10-8 B is a factor table for 5-gallon cans of joint compound. Figure 10-8 C is also a factor table for 5-gallon cans of joint compound. But use this table only if the ceiling is to be treated with a textured finish. Here's how to use these three tables:

In the first column, enter the total square feet of wallboard surface area. In our example, this comes to 10,016 SF. Then multiply the total area by the

Total surface area (square feet)	x	Factor	=	Pounds of nails
10,016	x	.00500	=	*50.08 or 50*

A 1⅜" annular ring nails

Total surface area (square feet)	x	Factor	=	Number of tubes
10,016	x	.00200	=	*20.03 or 20*

B Adhesive

Factors for wallboard fasteners
Figure 10-7

Total surface area (square feet)	x	Factor	=	Number of rolls
10,016	x	.00167	=	16.73 OR 17

A   Rolls of tape

Total surface area (square feet)	x	Factor	=	Number of cans
10,016	x	.00100	=	10.02 OR 10

B   Cans of joint compound

Total surface area (square feet)	x	Factor	=	Number of cans
10.016	x	.00250	=	25.04 OR 25

C   Cans of joint compound (for ceilings with textured finish)

Factors for wallboard finishing materials
Figure 10-8

factor to get the number of rolls of tape or cans of joint compound required.

Figure 10-8 A shows that we'll need 17 rolls of tape. Figure 10-8 B shows that 10 cans of joint compound are required. Figure 10-8 C shows that we'll need 25 cans of joint compound if we decide to treat the ceiling with a textured finish.

We've seen the easy way to calculate the materials required for insulation and wallboard. Now let's take a look at insulation and wallboard labor.

Insulation Labor

Insulation labor will vary, depending upon whether you install batt, blanket or loose fill insulation. The following tables are commonly used to estimate insulation labor.

Batt or blanket
R-value = 11 -----------180 square feet per manhour
R-value = 19 -----------140 square feet per manhour
R-value = 30 -----------120 square feet per manhour

Loose fill
Machine-blown ----------250 cubic feet per manhour
Hand-poured ------------ 20 cubic feet per manhour

Your insulation and wallboard estimates will be more accurate when they're based on your own

cost records. Look at Figures 10-9 and 10-10. These are weekly time sheets showing the insulation and wallboard installation on the Brown job.

Our daily logs show that it took a total of 68 manhours to install 7,199 SF of batt and blanket insulation in the ceilings, walls and floors on this job. Let's use this information to find the manhour factor for insulation.

Manhour factor for insulation— Look at Figure 10-11. In the first column, enter the total manhours required to install the insulation. In our example, 68 manhours were required to complete the insulation work. Now divide the total manhours (68) by the total square feet of insulation (7,199) to get the manhour factor (0.00945).

Use this manhour factor for labor estimates on similar jobs. For example, if you're estimating a job with 10,240 SF of insulation in the ceilings, walls and floors, here's how to estimate the labor required to do the job.

Multiply the factor (0.00945) by the total insulation area (10,240 SF) to get a total of 96.77 (97) manhours. This is the total labor required to do the job.

Wallboard Labor

Before wallboard application can begin, the carpenters have to do all the framing and furring. Make sure they've installed all the dropped ceilings for cabinets and all nailers for the inside corners on partitions running parallel to the ceiling joists. And they'll have to cut all openings in the walls and ceilings for bathroom accessories and access doors. This is referred to as preparation or coordination work for installing the drywall.

Look again at Figure 10-9. The daily log shows us that a total of 64 manhours were required to complete the preparation work for wallboard installation. Let's use this information to find the manhour factor for coordination work for installing wallboard.

Manhour factor for wallboard— Look at Figure 10-12. In the first column, enter the total manhours required for wallboard installation (64). In the next column, enter the total square feet of wallboard installed (10,016). Then divide the total manhours (64) by the total area (10,016 SF) to get the manhour factor (0.00639).

Use this manhour factor to estimate future jobs. Here's an example. If a job has an estimated 12,000 SF of gypsum wallboard, multiply the total area (12,000) by the factor (0.00639) to get a total of 76.68 (77) manhours. If the hourly wage rate is $25.00 per hour, the cost for coordination work for wallboard installation labor will be $1,925.00.

Weekly Time Sheet

For period ending 10-23-XX   Brown   job

	Name	Exemptions	October 18 M	19 T	20 W	21 T	22 F	23 S	Rate	Reg.	Over-time	Total earnings
			Days							**Hours worked**		
1	D.L.White		8	8	8	8	8	X		40		
2	J.R.Davis		8	8	8	8	8	X		40		
3												
4												
5												
6												
7												
8												
9												
10												
11												
12												
13												
14												
15												
16												
17												
18												
19												
20												

Daily Log

Monday   Preparation for wallboard installation (16 manhours)
Tuesday   Preparation for wallboard installation (16 manhours)
Wednesday Preparation for wallboard installation (16 manhours)
Thursday Preparation for wallboard installation (16 manhours)   10,016 SF of wallboard
Friday   Started installing insulation (16 manhours)
Saturday XXXX

Weekly time sheet
Figure 10-9

Weekly Time Sheet

For period ending _10-30-XX_   _Brown_ job

	Name	Exemptions	Days October 25 M	26 T	27 W	28 T	29 F	30 S	Rate	Hours worked Reg.	Over-time	Total earnings
1	D.L. White		8	8	8	8	8	X		40		
2	J.R. Davis		8	8	8	8	8	X		40		
3	W.R. Farlow		4	X	X	X	X	X		4		
4	D & A Plumbing Co.		✓	✓	✓	X	✓	X		—		
5												
6												
7												
8												
9												
10												
11												
12												
13												
14												
15												
16												
17												
18												
19												
20												

Daily Log

Monday _Insulation (20 manhours) --- Plumbing_
Tuesday _Insulation (16 manhours) --- Plumbing_
Wednesday _Finished insulation 7,199 SF of insulation (16 manhours) --- Plumbing_
Thursday _O/S Framing (16 manhours)_
Friday _O/S Framing (16 manhours)_
Saturday _XXXXX_

Weekly time sheet
Figure 10-10

Total manhours	Total ÷ insulation area = (square feet)		Manhour factor
68	÷	7,199 =	.00945

Insulation labor worksheet
Figure 10-11

Total manhours	Total ÷ wallboard area = (square feet)		Manhour factor
64	÷	10,016 =	.00639

Wallboard labor worksheet
Figure 10-12

Many builders subcontract the hanging and finishing of wallboard. Suppose you decide to accept a bid of $0.35 per square foot to hang and finish the wallboard. If there are 12,000 SF of wallboard, the cost will be $4,200.00.

Many drywall subs charge by the number of pieces installed. In this example, it would be 250 pieces. (There are 48 SF in each 4' x 12' panel.) Be sure to compare your labor costs before you decide to sub out your insulation and wallboard installation.

In this chapter we've seen how to make easy, accurate insulation and drywall estimates. Next, we'll take a look at siding, exterior and interior trim.

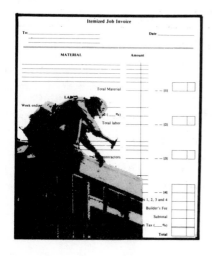

Chapter 11

Siding and Trim Records

Siding and trim, properly installed, serve two purposes. They improve the appearance of the house and protect the framework from weathering.

Siding comes in many styles, shapes, and patterns. It can be used alone (as the only cladding on a house), or in combination with brick veneer. You can also combine two or more types of siding, such as horizontal and vertical.

The exterior trim is usually installed at the same time as the rough-in for the electrical, plumbing, heating and air conditioning. The interior walls are also being finished at this time. Interior trim is the last carpentry work done in the house.

In this chapter, we'll look at the materials and labor required for siding, exterior trim and interior trim. We'll use our cost records as a guide to easy, accurate siding and trim estimates.

Siding Materials

To estimate any type of siding, first compute the wall area. Then add the gable area, if you're installing gable siding. Disregard any openings less than 50 SF.

Common types of siding include: wood bevel siding, tongue and groove siding, vertical board siding and panel siding. Here's how to estimate the quantity of siding and nails required to install each of these types of siding.

Board and Panel Siding

Figure 11-1 A is a factor table for different widths of wood bevel siding (lapped or rabbeted). Figure 11-1 B shows the factors for different widths of tongue and groove siding. Factors for vertical board siding are shown in Figure 11-1 C. For all three types of siding, factors are given for 8", 10" and 12" nominal widths. These factor tables allow for lapping and waste.

Here's how to use these three tables. Look down the first column and find the nominal width. In the next column, enter the net wall area (including gable area, when appropriate). Multiply the net wall area by the factor to get the total square feet of siding required to do the job. Sample calculations are shown in Figure 11-1 A, B and C.

Use Figure 11-1 D when you're installing 4' x 8' siding panels. Again, multiply the net wall area by the factor to get the total number of panels required.

Siding Nails

Figure 11-2 is a factor table for siding nails. Figure 11-2 A shows the nail factors for horizontal siding. Figure 11-2 B gives the factors for panel siding. Use Figure 11-2 C for cedar shake siding. Figure 11-2 D is for soffits and porch ceilings. And Figure 11-2 E is for calculating cornice nails. Here's how to use these tables:

Look down the first column and find the nail size. In the next column, enter the wall area (for horizontal siding, panel siding and cedar shake siding), the ceiling area (for soffits and porch ceilings), or the total linear feet of cornice. Multiply this number by the factor to get the number of pounds of nails required to do the job.

Nominal width of siding	Net wall area (square feet)	x Factor	= Square feet of siding
8"		x 1.34	=
10"	1,184	x 1.26	= 1,491.84
12"		x 1.21	=

A Wood bevel siding (lapped or rabbeted)

Nominal width of siding	Net wall area (square feet)	x Factor	= Square feet of siding
8"	1,184	x 1.37	= 1,622.08
10"		x 1.33	=
12"		x 1.31	=

B Tongue and groove siding

Nominal width of siding	Net wall area (square feet)	x Factor	= Square feet of siding
8"		x 1.10	=
10"		x 1.08	=
12"	1,184	x 1.07	= 1,266.88

C Vertical board siding

Panel size	Net wall area (square feet)	x Factor	= Number of panels
4' x 8'	1,184	x .03125	= 37

D Panel siding

Factors for siding
Figure 11-1

Nail size	Wall area (square feet)	x Factor	= Pounds of nails
6d		x .00600	=
8d	1,184	x .00700	= 8.29 or 9
10d		x .01000	=

A Horizontal siding

Nail size	Wall area (square feet)	x Factor	= Pounds of nails
6d	1,184	x .01700	= 20.13 or 21
8d		x .02100	=
10d		x .03000	=

B Panel siding

Nail size	Wall area (square feet)	x Factor	= Pounds of nails
6d		x .01000	=
8d	1,184	x .01500	= 17.76 or 18

C Cedar shake siding

Nail size	Ceiling area (square feet)	x Factor	= Pounds of nails
4d		x .01000	=
6d	832	x .01250	= 10.40 or 11

D Soffit and porch ceiling

Nail size	Cornice length (linear feet)	x Factor	= Pounds of nails
6d or 8d	550	x .01000	= 5.50 or 6

E Cornice

Factors for siding nails
Figure 11-2

Exterior Trim and Molding Materials

Exterior trim and molding components include: fascia, frieze and rake boards, and rust-resistant nails. Here's how to estimate these trim and molding components:

Type and size	Trim or molding length (linear feet)	x	Factor	x	Price per CLF	=	Cost of lumber
1"x 6" "B & Btr."	210	x	.01000	x	$54.85	=	$115.19
1"x 8" "B & Btr."	120	x	.01000	x	69.50	=	83.40
2¼" Ranch Type Trim	400	x	.01000	x	49.75	=	199.00
3¼" Base	340	x	.01000	x	71.50	=	243.10
½" x ¾" Oak Shoe	340	x	.01000	x	31.00	=	105.40
11/16" x 3 5/8" Crown Mold	88	x	.01000	x	88.25	=	77.66
		x	.01000	x		=	
		x	.01000	x		=	
		x	.01000	x		=	
		x	.01000	x		=	

Factors for exterior trim and molding
Figure 11-3

Fascia, Frieze and Rake Boards

Your blueprints and specs will tell you the size, type and grade of material to use. You'll also use the blueprints to compute the total linear feet of trim and molding required. Then add 5% for waste, and round up to the next multiple of 10.

Here's an example. Suppose the specs call for fascia, frieze and rake boards that are 1'' x 6''. The blueprints show there are 193 LF of fascia board required. Add 5% for waste (9.65 LF) to get 202.65 LF. Now round up to the next multiple of 10. You'll need a total of 210 LF of 1'' x 6'' fascia board.

Once you know the total linear feet of fascia required, finding the cost is easy. Look at Figure 11-3. In the first column, enter the type and size of trim (or molding) you plan to use. In the next column, enter the total linear feet you'll need to do the job. In our example, this comes to 210 LF. Multiply the total linear feet (210) by the factor (0.01000) to get 2.10. Then multiply 2.10 by the price per CLF ($54.85). This gives you a total cost of $115.19 for the fascia board on this job. Use the same table to compute the cost of each of the other trim and molding components.

Notice that the prices shown in this table are quoted per hundred linear feet (CLF). Use this table only when you need prices quoted per CLF.

Exterior Trim Nails

Use rust-resistant nails for all exterior trim and molding. Most trim and molding can be fastened with 6d or 8d nails. Here's how to compute the quantity of nails required:

Look again at Figure 11-2 E. Use this table to compute the quantity of nails required for exterior trim and molding. Just multiply the factor by the total linear feet of exterior trim or molding to get the number of pounds of nails required.

Let's compute the nails required for our 210 LF of fascia board. Multiply the total linear feet (210) by the factor (0.01000). We'll need 2.1 (or 3) lbs. of nails for this portion of the exterior trim and molding work. Use this same table to calculate the nails required for all the exterior trim and molding components.

Interior Trim and Molding Materials

When the wallboard is up and the walls have dried, the interior trim can begin. Interior trim is the last carpentry work done in a house. It includes flooring (when used in lieu of carpet), floor underlayment, baseboard, baseshoe, wall molding and paneling, interior doors (including trim and hardware), window trim, kitchen cabinets and countertops, vanities, bars, closet shelves, stairs, mirrors and medicine cabinets, tub and shower doors, decorative beams, and room dividers.

Length of room

Width of room	2'	3'	4'	5'	6'	7'	8'	9'	10'	11'	12'	13'	14'	15'	16'	17'	18'	19'	20'
2'	4	6	8	10	12	14	16	18	20	22	24	26	28	30	32	34	36	38	40
3'	6	9	12	15	18	21	24	27	30	33	36	39	42	45	48	51	54	57	60
4'	8	12	16	20	24	28	32	36	40	44	48	52	56	60	64	68	72	76	80
5'	10	15	20	25	30	35	40	45	50	55	60	65	70	75	80	85	90	95	100
6'	12	18	24	30	36	42	48	54	60	66	72	78	84	90	96	102	108	114	120
7'	14	21	28	35	42	49	56	63	70	77	84	91	98	105	112	119	126	133	140
8'	16	24	32	40	48	56	64	72	80	88	96	104	112	120	128	136	144	152	160
9'	18	27	36	45	54	63	72	81	90	99	108	117	126	135	144	153	162	171	180
10'	20	30	40	50	60	70	80	90	100	110	120	130	140	150	160	170	180	190	200
11'	22	33	44	55	66	77	88	99	110	121	132	143	154	165	176	187	198	209	220
12'	24	36	48	60	72	84	96	108	120	132	144	156	168	180	192	204	216	228	240
13'	26	39	52	65	78	91	104	117	130	143	156	169	182	195	208	221	234	247	260
14'	28	42	56	70	84	98	112	126	140	154	168	182	196	210	224	238	252	266	280
15'	30	45	60	75	90	105	120	135	150	165	180	195	210	225	240	255	270	285	300
16'	32	48	64	80	96	112	128	144	160	176	192	208	224	240	256	272	288	304	320
17'	34	51	68	85	102	119	136	153	170	187	204	221	238	255	272	289	306	323	340
18'	36	54	72	90	108	126	144	162	180	198	216	234	252	270	288	306	324	342	360
19'	38	57	76	95	114	133	152	171	190	209	228	247	266	285	304	323	342	361	380
20'	40	60	80	100	120	140	160	180	200	220	240	260	280	300	320	340	360	380	400

Total area (square feet) of floor or ceiling
Figure 11-4

We'll look at the most important interior trim components: flooring and floor underlayment, baseboard, baseshoe, wall molding, paneling and nails.

Flooring and Floor Underlayment

Look at Figure 11-4. Use this table to compute the total floor area for different-size rooms. For example, a room 12'0'' x 16'0'' has 192 SF of floor space. Once you know the area, you can calculate the material required for flooring and floor underlayment.

To find the total quantity of flooring or underlayment needed, add together the areas of each room, closet and hallway in the house. If you're installing 4' x 8' plywood panel underlayment, use the panel siding factor shown in Figure 11-1 D to compute the number of panels required. Here's an example:

A kitchen 12'0'' x 15'0'' has 180 SF of floor space. Multiply the net floor area (180 SF) by the factor (0.03125) to get a total of 5.63 (or 6) underlayment panels.

Here's another example. To cover a floor area of 2,460 SF with 4' x 8' particleboard, multiply the area (2,460 SF) by the factor (0.03125) to get 76.88 (or 77) particleboard panels.

Baseboard, Baseshoe, Molding and Paneling

Look at Figure 11-5. Use this chart to compute the perimeter of different-size rooms. For example, a room 12'0'' x 18'0'' has a perimeter of 60 LF. A closet 3'0'' x 7'0'' has a 20 LF perimeter. Once you know the perimeter of the room, you can order the material required for baseboard, baseshoe and molding. No door openings have been deducted in the table shown in Figure 11-5.

To compute the number of hardboard or plywood panels (4' wide) required, use the factor table shown in Figure 11-6. Just enter the room perimeter in the first column. Multiply the perimeter by the factor to get the number of panels required.

Interior Trim Nails

You'll need to calculate nail quantities for all of the following: floor underlayment, baseboard, baseshoe, and paneling. You'll also use adhesive to secure paneling in place.

Figure 11-7 is a factor table for underlayment, baseboard and baseshoe nails. Use Figure 11-7 A to compute the quantity of 6d ring-grooved nails needed for floor underlayment. Just multiply the floor area by the factor to get the number of pounds of nails required.

Width of room \ Length of room	2'	3'	4'	5'	6'	7'	8'	9'	10'	11'	12'	13'	14'	15'	16'	17'	18'	19'	20'
2'	8	10	12	14	16	18	20	22	24	26	28	30	32	34	36	38	40	42	44
3'	10	12	14	16	18	20	22	24	26	28	30	32	34	36	38	40	42	44	46
4'	12	14	16	18	20	22	24	26	28	30	32	34	36	38	40	42	44	46	48
5'	14	16	18	20	22	24	26	28	30	32	34	36	38	40	42	44	46	48	50
6'	16	18	20	22	24	26	28	30	32	34	36	38	40	42	44	46	48	50	52
7'	18	20	22	24	26	28	30	32	34	36	38	40	42	44	46	48	50	52	54
8'	20	22	24	26	28	30	32	34	36	38	40	42	44	46	48	50	52	54	56
9'	22	24	26	28	30	32	34	36	38	40	42	44	46	48	50	52	54	56	58
10'	24	26	28	30	32	34	36	38	40	42	44	46	48	50	52	54	56	58	60
11'	26	28	30	32	34	36	38	40	42	44	46	48	50	52	54	56	58	60	62
12'	28	30	32	34	36	38	40	42	44	46	48	50	52	54	56	58	60	62	64
13'	30	32	34	36	38	40	42	44	46	48	50	52	54	56	58	60	62	64	66
14'	32	34	36	38	40	42	44	46	48	50	52	54	56	58	60	62	64	66	68
15'	34	36	38	40	42	44	46	48	50	52	54	56	58	60	62	64	66	68	70
16'	36	38	40	42	44	46	48	50	52	54	56	58	60	62	64	66	68	70	72
17'	38	40	42	44	46	48	50	52	54	56	58	60	62	64	66	68	70	72	74
18'	40	42	44	46	48	50	52	54	56	58	60	62	64	66	68	70	72	74	76
19'	42	44	46	48	50	52	54	56	58	60	62	64	66	68	70	72	74	76	78
20'	44	46	48	50	52	54	56	58	60	62	64	66	68	70	72	74	76	78	80

Perimeter of room (linear feet)
Figure 11-5

Room perimeter (linear feet)	x	Factor	=	Number of panels
54	x	.25000	=	13.50 or 14

Factor for interior paneling (4' wide)
Figure 11-6

Use Figure 11-7 B to compute the quantity of 8d finish nails required for baseboard. Use Figure 11-7 C to figure the quantity of 6d finish nails required for baseshoe.

To compute the quantity of nails and adhesive you'll need for paneling, use Figure 11-8. Figure 11-8 A is for 1/4-lb. boxes of 1⅝'' colored nails. Figure 11-8 B is for pounds of 4d finish nails. Figure 11-8 C is for tubes of adhesive.

Siding and Exterior Trim Labor

Labor estimates for siding and exterior trim can be based either on manhours per unit of material installed or on manhours per square foot of floor space.

Floor area (square feet)	x	Factor	=	Pounds of nails
2,640	x	.02000	=	52.80 or 53

A Floor underlayment

Baseboard length (linear feet)	x	Factor	=	Pounds of nails
850	x	.01000	=	8.50 or 9

B Baseboard

Baseshoe length (linear feet)	x	Factor	=	Pounds of nails
250	x	.00500	=	1.25 or 2

C Baseshoe

Factors for underlayment, baseboard and baseshoe nails
Figure 11-7

Paneling area (square feet)	x	Factor	=	Boxes of colored nails
448	x	.01000	=	4.48 or 5

A 1⅝" colored nails

Paneling area (square feet)	x	Factor	=	Pounds of finish nails
448	x	.01000	=	4.48 or 5

B 4d finish nails

Paneling area (square feet)	x	Factor	=	Number of tubes
448	x	.01000	=	4.48 or 5

C Adhesive

Factors for paneling nails and adhesive
Figure 11-8

We'll figure it both ways in this chapter. Here are some figures to use if you base your estimate on units of material installed:

- Siding— 4-8 manhours per 100 SF of siding

- Cornice— 6-8 manhours per 100 LF of cornice

- Gable (rake)— 10-14 manhours per 100 LF of gable

- Other molding— 4-6 manhours per 100 LF of molding

These estimates may or may not be accurate for work done by your own crew. Your best labor estimate for any task will always come from the figures you keep in your own cost records.

Installing siding and trim on the second floor will require more time than installing them on the first floor. Why? Because second-floor work requires scaffolding that must be erected and dismantled.

Scaffolding labor is charged to exterior trim and siding. The labor required to install siding and trim on the second floor will generally be 25 to 35% higher than labor for the first floor.

Figures 11-9A through 11-9H are weekly time sheets for the Brown job. The daily logs show the labor required for siding, exterior trim and interior trim. The siding and most of the exterior trim were on the second floor. This means that scaffolding was needed and extra labor was required to put it up. The house was brick veneer with 4' x 8' panel siding on the second floor.

The daily logs show that a total of 263 manhours were required to install the second-floor siding and exterior trim on a house (and attached garage) with a total floor area of 2,850 SF. Let's use this information to find the manhour factor for siding and exterior trim.

Manhour factor for siding and exterior trim— Figure 11-10 is a labor worksheet for siding and exterior trim. In the first column, enter the total manhours. In our example, this comes to 263 manhours. Divide the total manhours by the total house and garage area (2,850 SF) to get a manhour factor of 0.09228. Now let's use the manhour factor to estimate a future job.

Let's say the new job we're bidding has a total house area of 1,875 SF. Multiply the house area (1,875 SF) by the manhour factor (0.09228) to get 173.03 (or 173) manhours.

Keep in mind that the manhour factor 0.09228 was developed on a job where most of the work was on the second floor. If all of the siding and exterior trim were on the first floor, the manhour factor would be reduced by about 25%. The manhour factor for first-floor siding and exterior trim is 0.06921.

If the new job we're bidding has the siding and exterior trim work only on the first floor, estimate the labor by multiplying the house area (1,875 SF) by the new factor (0.06921), to get a total of 129.77 (or 130) manhours required to do the job.

Interior Trim Labor

Use your weekly time sheets from previous jobs to estimate interior trim labor. Look again at Figures 11-9C through 11-9H. The daily logs show that a total of 411 manhours were required to install the interior trim for a house (and attached garage) with a total floor area of 2,850 SF. Now let's compute the manhour factor for interior trim.

Manhour factor for interior trim— Look at Figure 11-11. In the first column, enter the total manhours (411). Divide the total manhours by the total house and garage area (2,850 SF) to get a manhour factor of 0.14421.

Weekly Time Sheet

For period ending 3-12-XX   Brown   job

#	Name	Exemptions	March 7 M	8 T	9 W	10 T	11 F	12 S	Rate	Hours worked Reg.	Over-time	Total earnings
1	D.L. White		8	8	6½	8	8	X		38½		
2	A.L. King		X	8	8	8	8	X		32		
3	J.E. King		X	X	8	8	8	X		24		
4	Holston Drywall, Inc.		X	X	X	✓	X	X		—		
5												
6												
7												
8												
9												
10												
11												
12												
13												
14												
15												
16												
17												
18												
19												
20												

Daily Log

Monday Started waterproofing foundation

Tuesday Finished parging --- started bituminous coating

Wednesday Finished bituminous coating --- Drain tile around basement area

Thursday Drywall --- o/s trim (24 manhours)

Friday o/s trim (24 manhours)

Saturday XXXXX

Weekly time sheet
Figure 11-9A

Weekly Time Sheet

For period ending 3-19-XX   Brown   job

#	Name	Exemptions	Days March 14 M	15 T	16 W	17 T	18 F	19 S	Rate	Hours worked Reg.	Over-time	Total earnings
1	D.L. White		8	8	4	6	8	X		34		
2	A.L. King		8	8	8	8	8	X		40		
3	J.E. King		8	8	4	6	8	X		34		
4	Holston Drywall, Inc.		X	✓	X	✓	✓	X		—		
5												
6												
7												
8												
9												
10												
11												
12												
13												
14												
15												
16												
17												
18												
19												
20												

Daily Log

Monday Siding (24 manhours)
Tuesday Siding (24 manhours) ---Drywall finish
Wednesday Siding (16 manhours)
Thursday Siding---trim on back porch (20 manhours)---Drywall finish
Friday o/s trim (24 manhours)---Drywall finish
Saturday XXXXX

**Weekly time sheet
Figure 11-9B**

Weekly Time Sheet

For period ending 3-26-XX   Brown ___job

Name	Exemptions	Days March						Rate	Hours worked		Total earnings	
		21 M	22 T	23 W	24 T	25 F	26 S		Reg.	Over-time		
1 D.L.White		8	8	8	8	8	X		40			
2 A.L.King		8	8	8	8	8	X		40			
3 J.E.King		8	8	8	8	8	X		40			
4												
5												
6												
7												
8												
9												
10												
11												
12												
13												
14												
15												
16												
17												
18												
19												
20												

Daily Log

Monday Trim on front porch (18 manhours) --started floor underlayment (6 manhours)

Tuesday Floor underlayment (24 manhours)

Wednesday Floor underlayment (24 manhours)

Thursday Finished floor underlayment (42 manhours) --o/s trim (42 manhours)

Friday o/s trim (24 manhours)

Saturday XXXXX

Weely time sheet
Figure 11-9C

Weekly Time Sheet

For period ending 4-2-XX   Brown ____job

	Name	Exemptions	Days March 28 M	29 T	30 W	31 T	April 1 F	2 S	Rate	Hours worked Reg.	Over-time	Total earnings
1	D.L.White		8	8	8	8	8	X		40		
2	A.L.King		8	8	8	8	6½	X		38½		
3	J.E.King		8	8	8	8	6½	X		38½		
4												
5												
6												
7												
8												
9												
10												
11												
12												
13												
14												
15												
16												
17												
18												
19												
20												

Daily Log

Monday o/s trim (24 manhours)
Tuesday i/s trim (24 manhours)
Wednesday i/s trim (24 manhours)
Thursday i/s trim (24 manhours)
Friday i/s trim (21 manhours)
Saturday X X X X X

Weekly time sheet
Figure 11-9D

Weekly Time Sheet

For period ending 4-9-XX   Brown   job

	Name	Exemptions	\multicolumn{6}{c}{Days April}	Rate	Reg.	Over-time	Total earnings
			4 M	5 T	6 W	7 T	8 F	9 S				
1	D.L. White		8	8	8	8	8	X		40		
2	A.L. King		8	8	8	8	8	X		40		
3	J.E. King		8	8	8	8	8	X		40		
4	W.W. Peery, Paint contractor		✓	✓	✓	✓	X	X		—		
5												
6												
7												
8												
9												
10												
11												
12												
13												
14												
15												
16												
17												
18												
19												
20												

Hours worked

Daily Log

Monday i/s trim (24 manhours)---i/s painting

Tuesday i/s trim (24 manhours)---i/s painting

Wednesday i/s trim (24 manhours)---i/s painting

Thursday i/s trim (24 manhours)---i/s painting

Friday i/s trim (24 manhours)

Saturday XXXXX

Weekly time sheet
Figure 11-9E

Weekly Time Sheet

For period ending 4-16-XX   Brown   job

#	Name	Exemptions	Days April 11 M	12 T	13 W	14 T	15 F	16 S	Rate	Hours worked Reg.	Over-time	Total earnings
1	J. E. King		5	8	8	8	8	X		37		
2	W. W. Peery, Paint contractor		✓	X	✓	✓	X	X		—		
3	D & A Plumbing Co.		X	X	X	X	✓	X		—		
4												
5												
6												
7												
8												
9												
10												
11												
12												
13												
14												
15												
16												
17												
18												
19												
20												

Daily Log

Monday o/s trim (5 manhours)---o/s & i/s painting

Tuesday o/s trim (8 manhours)

Wednesday o/s trim (8 manhours)---i/s painting

Thursday o/s trim (8 manhours)---o/s painting

Friday Formed in for concrete (8 manhours)---plumbing

Saturday XXXXX

Weekly time sheet
Figure 11-9F

Weekly Time Sheet

For period ending 4-30-XX _____   _Brown_ _____ **job**

	Name	E x e m p t i o n s	Days April 25 M	26 T	27 W	28 T	29 F	30 S	Rate	Hours worked Reg.	Over-time	Total earnings	
1	D.L.White		X	6	8	X	3½	X		17½			
2	A.L.King		X	3½	8	8	8	X		27½			
3	J.E.King		X	6	8	X	3½	X		17½			
4	D.L.West		X	X	8	X	3½	X		11½			
5													
6													
7													
8													
9													
10													
11													
12													
13													
14													
15													
16													
17													
18													
19													
20													

Daily Log

Monday XXXXX

Tuesday Paneling in family room---i/s trim (15½ manhours)

Wednesday Paneling in family room---i/s trim (32 manhours)

Thursday Vanity cabinets in bathrooms (8 manhours)

Friday Finished paneling (18½ manhours)

Saturday XXXXX

Weekly time sheet
Figure 11-9G

Weekly Time Sheet

For period ending 5-7-XX   Brown _____ job

	Name	Exemptions	Days May						Rate	Hours worked		Total earnings
			2 M	3 T	4 W	5 T	6 F	7 S		Reg.	Over-time	
1	D.L. White		8	8	8	7	X	X		31		
2	A.L. King		8	8	8	7	X	X		31		
3	J.E. King		4	8	8	7	X	X		27		
4	D.L. West		4	8	8	7	X	X		27		
5	D & A Plumbing		X	X	✓	✓	X	X		—		
6	W.W. Peery, Paint contractor		X	X	✓	✓	✓	X		—		
7												
8												
9												
10												
11												
12												
13												
14												
15												
16												
17												
18												
19												
20												

Daily Log

Monday Trim in family room (16 manhours)   Forms for concrete (8 manhours)

Tuesday Kitchen cabinets (32 manhours)

Wednesday Finished Kitchen cabinets (16 manhours) Plumbing·· i/s painting --Cleaned brick (16 manhours)

Thursday Forms for concrete (18 manhours) --- Plumbing -- i/s painting cleaned brick (10 manhours)

Friday i/s painting

Saturday X X X X X

Weekly time sheet
Figure 11-9H

Total manhours	÷	House and garage area (square feet)	=	Manhour factor
263	÷	2,850	=	.09228

Labor worksheet for siding and exterior trim
Figure 11-10

Total manhours	÷	House and garage area (square feet)	=	Manhour factor
411	÷	2,850	=	.14421

Labor worksheet for interior trim
Figure 11-11

Let's use our interior trim manhour factor to bid on a new job. Suppose we're bidding on a house with a total floor area of 3,290 SF. Multiply the house area (3,290 SF) by the factor (0.14421) to get 474.45 (475) manhours. Use this as your interior trim manhour estimate for the new job.

In this chapter, we've learned how to make easy, accurate siding and trim estimates. In the next chapter, we'll use our cost records to estimate brickwork and concrete.

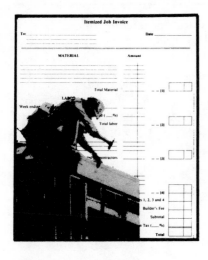

Chapter 12

Brickwork and Concrete Records

Most general contractors prefer to sub out their masonry work. Even if the general contractor supplies the brick or block, the subcontractor will hire the masons and helpers, and provide worker's compensation and liability insurance for his employees. The sub will also furnish all tools and necessary equipment.

Masonry subcontractors quote prices in two different ways. They can charge a set rate per thousand (M) brick or block, plus extra for fireplace, hearth, special walls and steps. Or they can bid the job at one price per the plans and specifications.

Either system works well so long as the subcontractor does quality work and leaves a clean job site when the work is finished. Have an understanding with your masonry sub on how mortar joints will be tooled. Also, get a certificate of insurance showing that he has liability and worker's compensation insurance.

In this chapter, we'll cover the materials and labor required for brickwork and concrete. And we'll set up a brickwork and concrete record system that will save you money on future jobs.

Brickwork Materials

When calculating the materials required for brickwork, you'll need to compute the number of bricks in all walls, chimneys, steps, and hearths. You'll also need to figure the amount of masonry cement and sand. And you may sometimes want to know in advance the number of courses of brick required.

Here's the easy way to calculate brickwork materials.

Brick for Walls

Figure 12-1 is a factor table for wall brick. Here's how to use it. Look down the first column and find the type of brick you plan to use. In the next column, enter the total wall area. Let's say the total wall area is 2,170 SF. Then multiply the wall area by the joint factor to find the number of bricks you'll need to do the job.

There are two sets of joint factors: one set of factors is for a 3/8'' joint and the other set of factors is for a 1/2'' joint. In our example, we'll use the factor for the 1/2'' joint (6.16). When we multiply the wall area (2,170 SF) by the joint factor (6.16), we get 13,367.20 bricks required to do the job.

These factors don't include any allowance for waste. Add 5% to your total quantity of bricks to allow for waste.

Brick for Chimneys

The chimney for the prefab fireplace is topped with brick. The brick extends from below the roof to 2'0'' above the ridge. This gives us a total chimney height of 5'0'' (5.0').

Type of Brick	Wall Area (square feet)	x	Factor 3/8" Joint	1/2" Joint	=	Number of Bricks
Standard brick 2¼" x 3¾" x 8"	2,170	x	6.55	(6.16)	=	13,367.20
Roman brick 1½" x 3¾" x 11½"		x	6.40	6.00	=	
Norman brick 2¼" x 3¾" x 11½"		x	4.60	4.40	=	
SCR brick 2¼" x 5½" x 11½"		x	---	4.40	=	

Factors for wall brick
12-1

Use the following table to compute the number of bricks required in chimneys. These factors are based on the number of bricks per foot of height for each flue size.

Number of flues	Size of flue	Factor
1	8" x 8"	27
1	8" x 12"	31
1	12" x 12"	35
2	8" x 8"	46
2	8" x 12"	55
2	12" x 12"	62

Let's try a sample calculation. If the flue is 12" x 12", how many bricks will you need? Multiply the chimney height (5.0') by the factor (35) to get a total of 175.0 bricks.

Brick for Steps

To compute the number of bricks needed for steps, multiply the number of steps by the number of bricks in each step. If you want a total of 10 steps, and each step has 58 bricks in it, multiply 10 by 58 to get a total of 580 bricks required for the steps.

Brick for Hearths

There are two factors for fireplace hearths. One factor is for standard brick laid flat (4.00000). The other factor is for standard brick laid on edge (6.34000).

Figure 12-2 is the factor table for fireplace hearths. In the first column, enter the area of the hearth (in square feet). In our example, the hearth area comes to 8 SF. Multiply the area by the factor. Then multiply this amount by the number of courses.

Hearth area (square feet)	x	Factor	x	Number of courses	=	Number of bricks
8	x	4.00000	x	2	=	64

A   Standard brick laid flat

Hearth area (square feet)	x	Factor	x	Number of courses	=	Number of bricks
8	x	6.34000	x	1	=	50.72

B   Standard brick laid on edge

Factors for hearth brick
Figure 12-2

Type of Brick	Number of bricks	x	Factor 3/8" joint	1/2" joint	=	Bags of cement
Standard brick 2¼" x 3¾" x 8"	15,000	x	.00466	(.00625)	=	93.75
Roman brick 1½" x 3¾" x 11½"		x	.00531	.00750	=	
Norman brick 2¼" x 3¾" x 11½"		x	.00587	.00818	=	
SCR brick 2¼" x 5½" x 11½"		x	---	.01148	=	

Factors for masonry cement
Figure 12-3

Figure 12-2 A tells us that 64 bricks are needed for the portion of the hearth where the bricks are laid flat. Figure 12-2 B shows that 50.72 bricks are required for the portion of the hearth where the bricks are laid on edge. Add these two together to get the total number of bricks required for the hearth. Then add on a 3% waste allowance.

Masonry Cement
Figure 12-3 is a factor table for masonry cement. Here's how to use it. Look down the first column and find the type of brick you plan to use. In the next column, enter the number of bricks in the job. Let's say it's a 15,000-brick job. Multiply the number of bricks by the joint factor to get the number of bags of cement needed to do the job.

Notice there are two sets of joint factors. One set of factors is for a 3/8" joint. The other set of factors is for a 1/2" joint. In our sample calculation, we'll use the joint factor for the 1/2" joint. Multiply the total number of bricks (15,000) by the joint factor (0.00625) to get a total of 93.75 bags of cement. Notice that the factors in this table include some allowance for waste.

Sand
Let's compute the number of tons of sand required for this 15,000-brick job. Figure 12-4 is a factor table for tons of sand.

Look down the first column and find the type of brick you're using. In the next column, enter the total number of bricks in the job (15,000). Multiply

Type of Brick	Number of bricks	x	Factor 3/8" Joint	1/2" Joint	=	Tons of sand (dry)
Standard brick 2¼" x 3¾" x 8"	15,000	x	.00047	(.00063)	=	9.45
Roman brick 1½" x 3¾" x 11½"		x	.00053	.00075	=	
Norman brick 2¼" x 3¾" x 11½"		x	.00059	.00082	=	
SCR brick 2¼" x 5½" x 11½"		x	---	.00115	=	

Factors for sand
Figure 12-4

Type of Brick	Height of walls	x	Factor 3/8" Joint	1/2" Joint	=	Number of brick courses
Standard brick 2¼" x 3¾" x 8"	3'0" (3.0')	x	4.57143	(4.36364)	=	13.09
Norman brick 2¼" x 3¾" x 11½"	7'4" (7.333')	x	4.57143	(4.36364)	=	32
SCR Brick 2¼" x 5½" x 11½"		x	4.57143	4.36364	=	
Roman brick 1½" x 3¾" x 11½"	11'3" (11.250')	x	(6.40000)	6.00000	=	72

Factors for courses of brick
Figure 12-5

the number of bricks by the joint factor for 1/2" joints (0.00063) to get a total of 9.45 tons of sand required for the job.

Notice that we have two sets of joint factors. One set of factors is for 3/8" joints. The other set of factors is for 1/2" joints. Make sure you use the right factor when computing the tons of sand required.

Notice also that the factors in this table are based on tons of *dry* sand. If the sand is wet, increase your sand estimate by 25% to 40%.

Number of Courses

Sometimes you'll want to know in advance how many courses of brick are needed. Figure 12-5 is a factor table for courses of brick. Here's how to use it. Look down the first column to find the type of brick. In the next column, enter the height of the wall. Let's say the wall height is 3'0". Multiply the wall height by the joint factor to get the number of courses required.

Again, we have two sets of joint factors. One set of factors is for 3/8" joints. The other set is for 1/2" joints. When we multiply the wall height (3.0') by the joint factor for 1/2" joints (4.36364), we get a total of 13.09 brick courses required. This wall is illustrated in Figure 12-6.

To convert feet and inches to decimal equivalents of a foot, refer to Figure 12-7. For example, the decimal equivalent of 4" is 0.333'. This means that 7'4" becomes 7.333'.

Concrete Materials

Concrete is the one material that's used in almost all new construction. For most general contractors, concrete is specialty work that is best handled by subcontractors. Specialists can do this work faster and cheaper than a crew who pours concrete only occasionally.

Concrete subcontractors normally charge by the square foot. Preparation work, such as installing forms, screeds, reinforcing rods and wire mesh, is in some areas, the responsibility of the general contractor. In this case, after the concrete is poured, it's the responsibility of the general contractor to remove the forms and screeds.

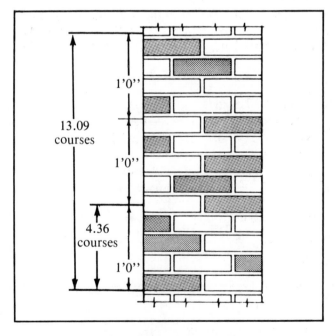

Standard brick with ½" joints
Figure 12-6

4th	8th	16th	0"	1"	2"	3"	4"	5"	6"	7"	8"	9"	10"	11"
		0	.000	.083	.167	.250	.333	.417	.500	.583	.667	.750	.833	.917
		1	.005	.089	.172	.255	.339	.422	.505	.589	.672	.755	.839	.922
	1	2	.010	.094	.177	.260	.344	.427	.510	.594	.677	.760	.844	.927
		3	.016	.099	.182	.266	.349	.432	.516	.599	.682	.766	.849	.932
1	2	4	.021	.104	.188	.271	.354	.438	.521	.604	.688	.771	.854	.938
		5	.026	.109	.193	.276	.359	.443	.526	.609	.693	.776	.859	.943
	3	6	.031	.115	.198	.281	.365	.448	.531	.615	.698	.781	.865	.948
		7	.036	.118	.203	.286	.370	.453	.536	.620	.703	.786	.870	.953
2	4	8	.042	.125	.208	.292	.375	.458	.542	.625	.708	.792	.875	.958
		9	.047	.130	.213	.297	.380	.464	.547	.630	.714	.797	.880	.964
	5	10	.052	.135	.219	.302	.386	.469	.552	.635	.719	.802	.885	.969
		11	.057	.141	.224	.307	.391	.474	.557	.641	.724	.807	.891	.974
3	6	12	.063	.146	.229	.313	.396	.479	.563	.646	.729	.813	.896	.979
		13	.068	.151	.234	.318	.401	.484	.568	.651	.734	.818	.901	.984
	7	14	.073	.156	.240	.323	.406	.490	.573	.656	.740	.823	.906	.989
		15	.078	.161	.245	.328	.411	.495	.578	.661	.745	.828	.911	.995

Decimal equivalents of fractional parts of a foot
Figure 12-7

Forms and screeds are usually constructed from extra lumber found on the job site. But you'll need to order the reinforcing rods, wire mesh, concrete and crushed stone. Here's how.

Reinforcing Rods
Use Figure 12-8 to compute the total linear feet of reinforcing rods required. Figure 12-8 A is for horizontal and transverse rods with 16" spacing.

Width _24.0'_ x .75 less 1 = _17_ x length _26.0'_ = _442_ linear feet

Length _26.0'_ x .75 less 1 = _18.5_ x width _24.0'_ = _444_ linear feet

OR 19 RODS

_886_ linear feet

Allow (_5_ %) for overlapping   _44.30_

Total   _930.30_ linear feet

Note: Order 940 LF rods

A Rods 16" x 16"

Width _____ less 1 = _____ x length _____ = _____ linear feet

Length _____ less 1 = _____ x width _____ = _____ linear feet

Allow (_____ %) for overlapping   _____

Total   _____ linear feet

B Rods 12" x 12"

Factors for reinforcing rods
Figure 12-8

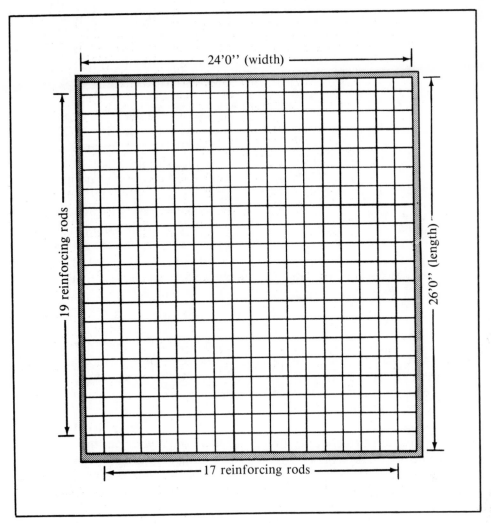

Reinforcing rods
(16'' x 16'')
Figure 12-9

Figure 12-8 B is for horizontal and transverse rods with 12'' spacing.

Let's try a sample calculation. Suppose you're using horizontal and transverse rods (16'' x 16'') for a concrete slab 24.0' wide by 26.0' long. How many linear feet of reinforcing rods will you need? Use the formula shown in Figure 8-12 A.

Multiply the width (24.0') by the factor (0.75), and subtract one. This comes to 17. Multiply 17 by the length (26.0') to get 442 linear feet.

Now multiply the length (26.0') by the factor (0.75), and subtract one. This comes to 18.5 (or 19 rods). Multiply 18.5 by the width (24.0') to get 444 linear feet.

Add together the two totals (442 and 444) to get 886 linear feet. Allow 5% for overlapping (44.30). This comes to 930.30 linear feet of reinforcing rods required. Order a total of 940 LF of rods.

Figure 12-9 shows the placement of these reinforcing rods with 16'' spacing. The 26.0' length will require 19 rods (rounded up from 18.5). The 24.0' width requires 17 rods.

If the reinforcing rods are sold by weight, Figure 12-10 will help you compute the cost of the rods. Multiply the total linear feet by the weight per foot to get the total weight of the rods. Then multiply the weight by the cost per pound.

If we're using 1/2'' number 4 rods for our sample slab, multiply the linear feet (940.0) by the weight per foot (0.668) to get a total weight of 627.92 lbs. Round this up to 628 lbs.

You'll also need tie wire for the rods. To find the number of pounds of tie wire required, just multiply the total linear feet of reinforcing rods by the tie-wire factor (0.00250). For our sample slab, multip-

| | Rod size | | Weight |
Diameter		Rod number	(pounds per foot)
¼''		2	.167
⅜''		3	.376
½''		4	.668
⅝''		5	1.043
¾''		6	1.502
⅞''		7	2.044
1''		8	2.670

Rod size and weight
Figure 12-10

ly the total linear feet (940.0) by the factor (0.00250) to get 2.35 (or 3) lbs. of tie wire required for the job.

Wire Mesh
Welded wire fabric or wire mesh is sold in rolls. Each roll is 5.0' x 150.0' (750 SF). To calculate the quantity of wire mesh required, multiply the total surface area by the wire-mesh factor (0.00133) to get the number of rolls.

For example, let's say you want to cover a surface area of 1,420 SF. Multiply the total area (1,420 SF) by the factor (0.00133). You'll need to order 1.89 (or 2) rolls of wire mesh for this job.

Concrete and Crushed Stone
Figure 12-11 is a factor table for concrete and crushed stone. Use these factors to simplify your concrete estimates. Here's how. Let's say we want to pour a slab 5'' thick in a garage 24'0'' x 24'0''. How many cubic yards of concrete will we need?

Look down the first column of Figure 12-11, and find the slab thickness (5''). The next column shows the factor for that thickness (0.01543). In the next column, enter the total area of the slab (in square feet). In our example, this comes to 576 SF. Multiply the factor (0.01543) by the area (576 SF) to get 8.89 cubic yards of concrete.

Here's another sample calculation. Suppose we want a 4'' slab covering an area of 1,420 SF. The factor for a 4'' slab is 0.01235. Multiply the factor by the area to get 17.54 cubic yards of concrete required to do the job.

If you plan to use 4'' of crushed stone under the concrete, just multiply the number of cubic yards (17.54) by the factor for crushed stone (1.35) to get the number of tons of crushed stone required. In this example, it comes to 23.68 tons of crushed stone.

Brickwork Labor
Even though your masonry subcontractor will provide tools, equipment, masons and helpers, you'll still have some important preparation and cleanup costs to include in your estimate.

Slab thickness	Factor x	Slab area (square feet) =	Cubic yards of concrete or crushed stone
3''	.00926 x		
3½''	.01080 x		
4''	.01235 x	*1,420*	*17.54*
4½''	.01389 x		
5''	.01543 x	*576*	*8.89*
5½''	.01698 x		
6''	.01852 x		

Cubic yards of crushed stone multiplied by 1.35 = tons of crushed stone.

(Note: Based on one cubic yard crushed stone = 2,700 pounds.)

Factors for concrete and crushed stone
Figure 12-11

Weekly Time Sheet

For period ending **5-7-XX**   **BROWN** job

	Name	Exemptions	Days MAY						Rate	Hours worked		Total earnings
			2 M	3 T	4 W	5 T	6 F	7 S		Reg.	Over-time	
1	D.L. WHITE		8	8	8	7	X	X		31		
2	4.L. KING		8	8	8	7	X	X		31		
3	J.E. KING		4	8	8	7	X	X		27		
4	D.L. WEST		4	8	8	7	X	X		27		
5	D&A PLUMBING		X	X	✓	✓	X	X		—		
6	W.W. PEERY, PAINT CONTRACTOR		X	X	✓	✓	✓	X		—		
7												
8												
9												
10												
11												
12												
13												
14												
15												
16												
17												
18												
19												
20												

Daily Log

Monday TRIM IN FAMILY ROOM (16 MANHOURS) -- FORMS FOR CONCRETE (8 MANHOURS)
Tuesday KITCHEN CABINETS (32 MANHOURS)
Wednesday FINISHED KITCHEN CABINETS (16 MANHOURS) -- CLEANED BRICK (16 MANHOURS) PLUMBING ··· 1/S PAINTING
Thursday FORMS FOR CONCRETE (18 MANHOURS) -- CLEANED BRICK (10 MANHOURS) PLUMBING ··· 1/S PAINTING
Friday 1/S PAINTING
Saturday XXXXX

Weekly time sheet
Figure 12-12A

Weekly Time Sheet

For period ending ___4-16-XX___   ___BROWN___ job

| | Name | Exemptions | Days APRIL | | | | | | Rate | Hours worked | | Total earnings |
			11 M	12 T	13 W	14 T	15 F	16 S		Reg.	Over-time	
1	J.E. KING		5	8	8	8	8	X		37		
2	W.W. PEERY, PAINT CONTRACTOR		✓	X	✓	✓	X	X		—		
3	D.& A PLUMBING CO.		X	X	X	X	✓	X		—		
4												
5												
6												
7												
8												
9												
10												
11												
12												
13												
14												
15												
16												
17												
18												
19												
20												

Daily Log

Monday O/S TRIM (5 MAN-HOURS) --- O/S & I/S PAINTING

Tuesday O/S TRIM (8 MAN-HOURS)

Wednesday O/S TRIM (8 MAN-HOURS) I/S PAINTING

Thursday O/S TRIM (8 MAN-HOURS) --- O/S PAINTING

Friday FORMED IN FOR CONCRETE (8 MAN-HOURS) --- PAINTING

Saturday X X X X X

Weekly time sheet
Figure 12-12B

Weekly Time Sheet

For period ending 5-14-XX   BROWN _job

	Name	Exemptions	Days MAY 9 M	10 T	11 W	12 T	13 F	14 S	Rate	Hours worked Reg.	Over-time	Total earnings
1	D.L. WHITE		8	8	8	X	X	X		24		
2	D.L. WEST		8	8	8	X	X	X		24		
3	WALTON CONSTRUCTION		✓	X	X	✓	X	X		—		
4	A.L. KING		X	8	8	X	X	X		16		
5	J.E. KING		X	8	8	X	X	X		16		
6	W.W. PEERY, PAINT CONTRACTOR		X	X	X	X	✓	X		—		
7												
8												
9												
10												
11												
12												
13												
14												
15												
16												
17												
18												
19												
20												

Daily Log

Monday CONCRETE SCREEDS IN BASEMENT (16 MANHOURS)··· POURED CONCRETE IN GARAGE & BACK PORCH

Tuesday FORMED IN FRONT PORCH & WALK (16 MANHOURS)··· CARPENTRY WORK ON BACK PORCH (16 MANHOURS)

Wednesday PLACED REINFORCING RODS (16 MANHOURS)·· CARPENTRY WORK ON BACK PORCH (16 MANHOURS)

Thursday POURED CONCRETE IN BASEMENT, FRONT PORCH AND WALK

Friday O/S PAINTING

Saturday XXXXX

Weekly time sheet
Figure 12-12C

For example, your carpenters will have to frame an opening for the chimney and frame around the prefab fireplace (if there is one). If the masonry subcontractor doesn't include cleaning the brick in his contract, your carpenter helpers will probably have to do it. Don't overlook these important costs.

Figures 12-12A, 12-12B and 12-12C are weekly time sheets for the Brown job. Figure 12-12A shows that 26 manhours were spent cleaning 15,000 bricks on the Brown job. At an hourly wage rate of $25.00 per hour, this is an additional $650.00 that should be included the masonry estimate.

Manhour factor for masonry preparation and cleanup— Figure 12-13 is a factor table for masonry preparation and cleanup work. Here's how to use it. Let's say that 10 manhours are required to cut an opening for the chimney and put up framing for the prefab fireplace. Add this to the 26 manhours required for cleaning the brick. This gives us a total of 36 manhours required for masonry prep and cleanup on this job.

Total manhours	÷	Total number of bricks	=	Manhour factor
36	÷	15,000	=	.00240

Factor for masonry prep and cleanup
Figure 12-13

In the first column of Figure 12-13, enter the total manhours (36). In the next column enter the total number of bricks (15,000). Then divide the number of manhours by the number of bricks to get a manhour factor of 0.00240. Use this manhour factor to estimate the labor for masonry prep and cleanup work on future jobs.

Here's an example. Suppose you need 22,500 bricks for a house that you're bidding on. Multiply the manhour factor (0.00240) by the total number of bricks (22,500) to get 54 manhours required to do the prep and cleanup work. Multiply the total manhours (54) by the hourly wage rate ($25.00 per hour) to get the estimated cost of the preparation

and cleanup work. In this example, it comes to $1,350.00.

Concrete Labor

Now let's compute the manhours required for concrete work. Look again at the weekly time sheets for the Brown job, Figures 12-12A, 12-12B and 12-12C.

Figure 12-12A shows that 26 manhours were spent installing forms for the concrete. Figure 12-12B shows another 8 hours spent on formwork. Figure 12-12C tells us that an additional 48 hours were spent on concrete work. This time was spent placing the screeds, forms and reinforcing rods. This comes to a total of 82 manhours spent on concrete labor covering an area of 1,989 SF.

Manhour factor for forms, screeds and reinforcing rods— Use Figure 12-14 to compute the manhour factor for concrete forms, screeds and reinforcing rods. Here's a sample calculation.

Total manhours	÷	Total area of concrete (square feet)	=	Manhour factor
82	÷	1,989	=	.04123

Factor for concrete forms, screeds and reinforcing rods
Figure 12-14

In the first column, enter the total manhours. In our example, this comes to 82 manhours. In the next column, enter the total area (1,989 SF). Divide the total manhours by the total area to get a manhour factor of 0.04123. Use this manhour factor for future bids.

Let's say you're bidding a job with 1,685 SF of concrete. Multiply the total area (1,685 SF) by the manhour factor (0.04123) to get a total of 69.47, or 70 manhours. If the hourly labor rate is $25.00 per hour, your estimated labor cost for the concrete forms on this job will be $1,750.00.

In this chapter, we've learned how to make easy, accurate brickwork and concrete estimates. In the next chapter, we'll see how to reduce our costs for subcontracted work.

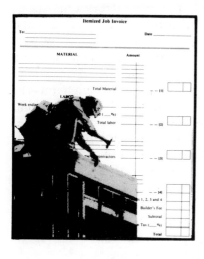

Chapter 13

Subcontracted Work Records

There's a major advantage to subcontracting part of a construction project. The subcontractor gives you a firm bid and a set price. The total cost is known before work begins. You, the general contractor, are protected. You rely on the sub's estimate and make it a part of your bid.

In theory, the subcontractor accepts the risk of loss if costs exceed his estimate. The work should be done for the price quoted and no more.

But experienced builders agree it isn't always that simple. Yes, the subcontractor's quote was firm. But it was based on a particular set of plans and job description. Now he claims that you're asking him to do something that wasn't in the plans and wasn't included in his bid. That will cost more!

You don't see it that way, of course. But as the general contractor, you have little choice. The work has to be done or you don't get paid.

And you may not be able to pass this extra cost along to the owner. The plans assume a complete building that complies with the building code and is ready for occupancy. The owner will be unwilling to pay extra for what he considers was included in your bid, even if the subcontractor insists it wasn't in *his* bid. The result: You're stuck with extra costs that reduce your profit.

In this chapter, we'll learn how to avoid common subcontract problems. We'll set up a basic subcontract that will reduce your risk of loss. We'll also set up the important specialty subcontracts.

These include: excavation and fill, masonry, roofing, electrical, plumbing, heating, and air conditioning. Finally, we'll review our cost records for subcontracted work.

Basic Subcontract

Experienced contractors are usually quick to spot potential problems with subcontracts. Once you've had a dispute on some point with a subcontractor, you're likely to anticipate similar problems in the future. But that's an expensive way to learn contracting and estimating.

How can you be sure that a subcontractor's bid will cover all of the work you expect to have done? Keep a checklist of the common items that run up subcontract costs in residential construction. Review this checklist before accepting a bid.

Common subcontract problems include: insurance and taxes, owner-subcontractor agreements, subcontractor pay period, and qualified bids. Let's look at how to avoid each of these problems. We'll also take a look at how to handle the unexpected problems.

Proof of insurance and taxes— Subcontractors operate their own businesses, hire their own employees and are responsible for FICA and FUTA taxes, worker's compensation, and liability insurance for their employees.

But you can still be held responsible for a sub's unpaid employee taxes or for an injury to or by a subcontractor's tradesmen. That's why you should always have proof of insurance coverage and deal only with reputable subs that comply with the tax laws.

Owner-subcontractor agreements— Sometimes the owner will enter into a separate agreement with one or more of the subcontractors on the job. This can put you in a difficult position.

You have no authority over that sub's tradesmen, but have to share space with them and coordinate the work they do. Have a written understanding with the owner stating who is liable for personal injury or property damage caused by that subcontractor or his employees.

Subcontractor pay period— Make sure every subcontract states clearly when the subcontractor will be paid. Will he receive a single payment when his work is finished? Or will he be paid as each phase is complete? Be clear on this, and *get it in writing.*

Bids "per plans and specs"— Get subcontract bids "per the owner's plans and specs" and with no other qualifications, if possible. The chance of a dispute is greater when the sub doesn't bid "per plans and specs."

A subcontract based on plans drawn up by the sub or including an ambiguous job description can mean trouble for the general contractor. The more descriptions, explanations and qualifications the subcontractor submits with his bid, the more likely the general contractor is to get stuck with additional costs.

If the sub doesn't bid per the owner's plans and specs, the sub is required to do only what he proposes and no more. If the sub omits something that's in the owner's plans, it will be your responsibility to correct it. That puts you in the middle again. And that can be a very expensive position.

Electrical and HVAC subs routinely draw up their own plans. That's fine as long as the subcontractor's plans *become the owner's plans* and are submitted as part of your bid. Make the sub's qualifications part of your contract with the owner. That will keep you out of some disputes and save you more than just a few dollars.

Unexpected problems— Even the most complete, professional plans and specs may omit some important point, have inconsistent provisions or leave room for misunderstanding. It's not necessarily the architect's fault. Detailed planning of even a simple home is a complex and demanding task.

A few changes and some improvising are inevitable on nearly every project. And the bigger the job, the more changes and improvising will be needed. That's just the nature of construction.

When subcontracted work costs more than you allowed for in your bid, make a detailed record of the problems, how they could've been avoided and how you resolved them. Document the claims of the subcontractor, and prepare your case for extra compensation from the owner.

Keep a correspondence file and make handwritten summaries of phone calls or conversations that relate to the increased costs. These notes that you make to yourself will be admissible in court if the dispute becomes a legal matter. The more complete your mastery of the facts, the more likely you are to recover the full amount of the subcontractor's additional charges.

Specialty Subcontracts

Let's look at some points to remember before accepting bids from the important specialty subcontractors: excavation and fill, masonry, roofing, electrical, plumbing, heating, and air conditioning.

Excavation and Fill

On most jobs, the first subcontractor is the excavation sub. Be sure to ask the following questions when setting up your subcontract:

1) Are the grade elevations specified for cuts and fills? If fill dirt is necessary, is the method of compaction specified?

2) Will it cost extra to move excavation equipment onto the job site? If so, how much will it cost?

3) If rock removal is required, or if there is disturbed earth (such as an underground spring), is the extra cost specified in the contract? Will this extra work be charged by the contract or by the hour? How much will it cost?

4) Does the contract require the removal of trees, stumps, debris, old buildings and fences? If so, how much will it cost?

5) Is the building site large enough to store the excavated soil during the construction period? Or will the soil have to be hauled away? Does the contract include the cost of this hauling? Does it specify the cost of hauling the soil back to the job site if this is necessary?

6) Will there be an extra charge for equipment during cold and wet weather?

7) If a temporary access road is necessary to deliver material to the job site, is this work included in the contract? If so, who will spread the gravel?

Masonry

Foundation block is priced either per block or by the job. Brick masons normally charge a fixed fee per thousand (M) bricks, plus extra for fireplaces, hearths, special walls, and steps.

Make sure that you and the masons have a clear understanding of the following points:

1) How many courses of block are required?

2) Who is responsible for leveling the foundation walls? How will they be leveled?

3) If brick is to be set on the block, where does the brick start (on which course of block)?

4) What is the spacing for anchor bolts?

5) On which block course will you set lintels for the foundation windows and doors?

6) Verify the location of all corners and their dimensions. Verify the size of block in each section of the foundation.

7) Check the size and location of windows, vents and doors in the foundation wall. A window described as 3'2'' wide can easily be misread as 32''. Also verify the size and location of notches for beams, girders or floor joists.

8) Check the location and size of all piers. Verify how they'll be constructed.

9) Know where you'll be using solid masonry blocks.

10) Verify the location and spacing of wall ties.

11) Know where and how mortar joints will be tooled.

12) Verify the correct mix for mortar.

13) In cold weather, what precautions will the masons take to keep mortar from freezing?

The time you spend going over these items with the masonry subcontractor is small compared to the cost of correcting mistakes later.

Roofing

Roofing subcontractors normally charge by the square (100 SF). The cost per square depends on the roofing material, roof pitch, and the number of valleys, vents and metal drip caps. Make it clear whether flashing is to be installed by the carpenters or the roofers.

Verify insurance coverage before signing a subcontract with the roofer. Worker's compensation insurance for roofers is one of the major costs for a roofing subcontractor.

Get the roofing sub to sign a complete specification sheet as part of the subcontract. Specs should include:

1) Felt paper: weight

2) Metal drip edging

3) Roof vents: type and size

4) Flashing: type and size, where used, how installed. Are valleys open, closed or woven? Who applies the roof cement?

5) Roof shingles: type and weight

6) Nails: size and number per shingle

7) Insurance: If the roofing subcontractor doesn't have worker's compensation and liability insurance for his employees, you must provide it.

Electrical

The electrical subcontractor normally bids on labor and materials for the rough-in electrical work.

Make sure the electrical sub understands where his responsibility begins. It normally begins at the meter. If the electric service line to the house is underground, the utility company will probably charge for the run to the meter. This can be a major cost item, and it probably isn't covered in the electrical subcontractor's bid.

Use the following checklist for your electrical subcontract:

1) Cost of service line to house

2) Cost of hookup for temporary electric service

3) Labor and materials for rough-in. This includes: wiring, outlets, plates, boxes for the fixtures, switches, connectors, and entrance panel circuit breakers.

4) Cost of light fixtures (if furnished by the electrical subcontractor)

5) Installation of light fixtures

6) Hookup for appliances

7) Hookup for the heating, cooling and ventilating equipment

8) Hookup and installation of any special electrical equipment

9) Telephone boxes and service to house

10) Television boxes and service to house

Plumbing

Most areas require a license to do plumbing work. And, of course, all plumbing must comply with the building and plumbing codes in each community.

An inspector will check the work for compliance. If the house is in an area where there's no public water service or public sewer, the local health department will probably have authority to approve the water and sewer system plans.

Estimating materials for the plumbing rough-in is work for a specialist. It should be done by the plumbing subcontractor. Plumbing fixtures, such as the kitchen sink, garbage disposal, bathtubs, shower stalls, water closets and lavatories are normally selected by the owner and installed by the plumber.

Review the following points before signing an agreement with the plumbing subcontractor:

1) Cost of water supply line from point of connection to the house, including trench for the water pipe.

2) Temporary water service hookup.

3) Cost of sewer line from the house to point of discharge, including trench for the sewer pipe.

4) Cost of labor and materials for the rough-in. This includes: type and size of water supply pipes, drain pipes, vent pipes, number and location of outside water faucets, and all materials required for connections.

5) Cost of plumbing fixtures, if furnished by the plumbing subcontractor.

6) Installation of plumbing fixtures, if furnished by the owner.

7) Installation of special equipment, such as water softeners, pressure valves, sump pumps or shower doors.

Heating and Air Conditioning

Many types of heating and cooling systems are available today: oil, gas, electric, space heaters, electric radiant heat with heating cables in the ceiling, heat pumps that are also air conditioners, and solar heaters.

Your heating system should be designed and installed by a reputable heating subcontractor. The system must meet all code requirements. If the unit is too small, too large or isn't properly installed, it won't operate efficiently.

Air conditioning units must be safe, quiet and economical. The capacity of the system depends on the size of the house and the calculated heat loss. The system must meet all building code requirements. All central air conditioning should be designed and installed by a reputable air-conditioning subcontractor.

If separate heating and air-conditioning systems are planned for a new home, consider using a heat pump. The heat pump has a dual function: It provides heat in cold weather and cools the house in hot weather. And installing a heat pump will cost less than installing separate heating and air-conditioning systems.

Verify these important points with your HVAC subcontractor:

1) Does the heat pump have an auxiliary resistance heating system that will automatically switch on in very cold weather?

2) Who constructs the concrete pad for the compressor?

Miscellaneous Subcontracts

You may also use subcontractors for: insulation, plaster or drywall, concrete finishing, painting, floor and wall covering, gutters, paving, and landscaping. Here are points to remember about agreements with each of these subs:

Insulation— The insulation sub's bid will be for labor and materials per plans and specs.

Plaster or drywall— The plaster sub will bid on labor and materials for the entire job, or he'll quote a fixed fee (per square yard) for labor only. The drywall sub normally bids only on the cost of labor.

Concrete— These subs normally charge a fixed fee per square foot. They pour and finish the concrete *after* forms and screeds are in place.

Paint— The bid will be for labor and materials per the job plans and specs. The bid can include both interior and exterior work.

Floor and wall covering— You may want to sub out the installation of wall-to-wall carpet, resilient floor tile, masonry floor covering, wall tile, and wall fabric. The owner normally selects the materials. The subcontract is for labor only.

Gutters— These bids will be for labor and materials per plans and specs.

Paving and landscaping— These bids will also be for labor and materials per plans and specs.

Recording Subcontract Costs

Figure 3-3, in Chapter 3, shows subcontract cost records for the Baker, Brown and Green jobs. The entries shown in Figure 3-3 are only for subcontract agreements with the general contractor. These entries don't include any owner-subcontractor agreements.

Be aware of the cost of work done by owner-subcontractor agreement. These costs may help you estimate future jobs.

Look again at Figure 3-3, in Chapter 3. Column 1 shows the total disbursement paid to each subcontractor. The disbursement is then charged to the appropriate jobs in columns 2, 3 and 4. The total shown in column 1 (on each line) must always equal the total of columns 2, 3 and 4 (on each line).

Look at the running totals through the month of June:

Baker job	$1,316.33
Brown job	1,840.32
Green job	2,780.00
Total	**$5,936.65**

The running total shown in column 1 ($5,936.65) matches the combined running totals shown in columns 2, 3 and 4.

Figure 3-3 shows your current subcontract costs for any job listed. It shows you the amount paid to each subcontractor on each job. Let's look at the cost breakdown for the Green job through September 30.

Masonry (June 2)	$2,780.00
Roofing (August 12)	544.30
Masonry (September 23)	919.69
Masonry (September 30)	683.85
Total	**$4,927.84**

Keep a separate ledger of subcontracted work for each fiscal year. A job that continues from one tax year to the next will appear on more than one ledger. This makes figuring costs a little more difficult. But it's essential to have separate ledgers for tax purposes.

In this chapter, we've learned how to set up subcontracts that will protect our profits. In the next chapter, we'll use our cost record system to make a composite labor estimate. We'll also review the important points of our complete cost recording and estimating system.

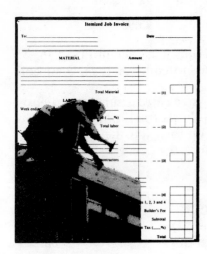

Chapter 14

Composite Labor Records

Every successful builder has built his business on quality work, integrity and creditability, accurate cost records, and accurate estimates.

We haven't covered quality and integrity in this book. But few builders survive for many seasons if their work doesn't reflect quality craftsmanship, or if they aren't honest and fair in their dealings.

We've seen how easy it is to keep accurate cost records and how to use these records to make accurate estimates.

Remember, there are two profits in every job. The money that goes into your pocket is only one. The second is what you learn on the job. Your cost records will provide you with valuable information again and again for as long as you're in the construction business.

In this chapter, we'll look at the last important cost record compiled for every job: the composite labor record. Then we'll review the importance of keeping accurate cost records.

Composite Labor Record

The last cost record compiled for every job is a manhour factor for the entire project. Figure 14-1 is a worksheet for computing this important factor. It shows how many manhours were needed per square foot of floor area.

This number will vary from job to job. The more work you subcontract, the lower this manhour factor will be.

Use this factor to make your preliminary labor estimates on similar jobs. And use it as a quick check against your final estimates. If it helps you spot a major error in addition or a key omission on a bid, it's worth far more than the few minutes the computation takes.

Here's how to use the composite labor worksheet shown in Figure 14-1. Let's figure the composite manhour factor for a house with an attached garage and a total area of 3,785 SF. The weekly time sheets for the entire job show that 2,863 manhours were required to complete the work on the house.

Divide the total manhours (2,863) by the total area (3,785 SF) to get a manhour factor of 0.75641. Now use this manhour factor to estimate the labor costs on a similar job.

Suppose a house has an attached garage and a total area of 2,850 SF. Multiply the total area

Total manhours	÷	Total area of house and garage (Square feet)	=	Manhour factor
2,863	÷	3,785	=	.75641

Composite labor worksheet
Figure 14-1

(2,850 SF) by the manhour factor (0.75641). This gives us a total of 2,155.77 (or 2,156) manhours required to do the job. Multiply the total manhours (2,156) by the hourly wage rate ($25.00) to get a total labor estimate of $53,900.00.

Keeping Accurate Cost Records

Accurate estimates separate professional builders from novices. If you bid too high, you lose the job. If you bid too low, you're not running a business. You're running an expensive hobby.

Your cost records are your source for accurate estimates. Most builders who fail in business keep poor records or no records at all. That's an established fact. But did they have poor records because they were failing, or did they fail because they had poor records?

The answer may be partially both. But it's clear that successful construction companies keep accurate cost records. If you want to build a successful construction company, be as good at record-keeping as you are at building and estimating.

Let's review why your cost records are so valuable:

1) Cost records provide a continuous breakdown of expenses during construction on each job.

2) Accurate cost records show you at once if any area of the project is exceeding your estimate.

3) These records provide essential data for estimates on future jobs.

4) Good cost records show wasted materials, inefficient or excess labor, padded payrolls and lost time.

Forms

Many of the forms used in this book are available at stationery stores. But some of the forms were custom-designed for builders. Blank copies of the custom-designed forms are provided in the following pages. Use your office copier or have an "instant printer" run off several hundred copies and make them into a pad.

Wrapping It All Up

I hope you're convinced that record-keeping is both essential to your survival as a construction contractor and a fairly simple task that yields big dividends. Good cost records are the raw material for developing quicker, easier, more accurate estimates.

But accurate records and estimates alone won't make you a successful builder. You have to add the remaining ingredients: quality work, integrity and creditability. Knowing how to compile and use cost records gives you an advantage over every builder who wastes this valuable information or doesn't use the data he has available.

Site Preparation

Date: _____   _____ job

	Estimated Costs	Actual Costs
1) Preliminary work:		
Architect's fee or stock plans	$_____	$_____
Plot plans	$_____	$_____
Building permit	$_____	$_____
Temporary water service	$_____	$_____
Sewer connection	$_____	$_____
Temporary phone and electric service	$_____	$_____
2) Site clearing	$_____	$_____
3) Excavation	$_____	$_____

Note: Actual costs $_____ divided
by _____ cu. yds. = $_____ per cu. yd.

Blasting required: (Yes) - (No)

Weather conditions: (Good) - (Fair) - (Poor)

	Estimated Costs	Actual Costs
4) Fill dirt _____	$_____	$_____

Note: Actual cost $_____ divided by _____
cu. yd. = $_____ per cu. yd.

	Estimated Costs	Actual Costs
5) Site cleaning and hauling	$_____	$_____

Itemized Job Invoice

To:_____   Date _____
_____
_____

MATERIAL   **Amount**

_____   _____

Total Material _____   — — [1]

LABOR

Week ending: _____

Payroll taxes, insurance and overhead (____%)   _____

Total labor _____   — — [2]

SUBCONTRACTORS

_____   _____

Total Subcontractors _____   — — [3]

OTHER CHARGES

_____   _____

Total Other Charges _____   — — [4]

Total of lines 1, 2, 3 and 4

Builder's Fee

Subtotal

Business and Occupation Tax (____%)

Total

Weekly Time Sheet

For period ending _____   _____job

	Name	Exemptions	Days						Rate	Hours worked		Total earnings	
			M	T	W	T	F	S		Reg.	Over-time		
1													
2													
3													
4													
5													
6													
7													
8													
9													
10													
11													
12													
13													
14													
15													
16													
17													
18													
19													
20													

Daily Log

Monday _____

Tuesday _____

Wednesday _____

Thursday _____

Friday _____

Saturday _____

Index

Other Practical References

National Construction Estimator

Current building costs in dollars and cents for residential, commercial and industrial construction. Prices for every commonly used building material, and the proper labor cost associated with installation of the material. Everything figured out to give you the "in place" cost in seconds. Many time-saving rules of thumb, waste and coverage factors and estimating tables are included. **528 pages, 8½ x 11, $18.50. Revised annually.**

Berger Building Cost File

Labor and material costs needed to estimate major projects: shopping centers and stores, hospitals, educational facilities, office complexes, industrial and institutional buildings, and housing projects. All cost estimates show both the manhours required and the typical crew needed so you can figure the price and schedule the work quickly and easily. **288 pages, 8½ x 11, $30.00**

Building Cost Manual

Square foot costs for residential, commercial, industrial, and farm buildings. In a few minutes you work up a reliable budget estimate based on the actual materials and design features, area, shape, wall height, number of floors and support requirements. Most important, you include all the important variables that can make any building unique from a cost standpoint. **240 pages, 8½ x 11, $14.00. Revised annually**

Construction Estimating Reference Data

Collected in this single volume are the building estimator's 300 most useful estimating reference tables. Labor requirements for nearly every type of construction are included: site work, concrete work, masonry, steel, carpentry, thermal & moisture protection, doors and windows, finishes, mechanical and electrical. Each section explains in detail the work being estimated and gives the appropriate crew size and equipment needed. **368 pages, 11 x 8½, $20.00**

Estimating Home Building Costs

Estimate every phase of residential construction from site costs to the profit margin you should include in your bid. Shows how to keep track of manhours and make accurate labor cost estimates for footings, foundations, framing and sheathing finishes, electrical, plumbing and more. Explains the work being estimated and provides sample cost estimate worksheets with complete instructions for each job phase. **320 pages, 5½ x 8½, $17.00**

Drywall Contracting

How to do professional quality drywall work, how to plan and estimate each job, and how to start and keep your drywall business thriving. Covers the eight essential steps in making any drywall estimate, how to achieve the six most commonly-used surface treatments, how to work with metal studs, and how to solve and prevent most common drywall problems. **288 pages, 5½ x 8½, $18.25**

Contractor's Guide to the Building Code

Explains in plain English exactly what the Uniform Building Code requires and shows how to design and construct residential and light commercial buildings that will pass inspection the first time. Suggests how to work with the inspector to minimize construction costs, what common building short cuts are likely to be cited, and where exceptions are granted. **312 pages, 5½ x 8½, $16.25**

Estimating Plumbing Costs

Offers a basic procedure for estimating materials, labor, and direct and indirect costs for residential and commercial plumbing jobs. Explains how to interpret and understand plot plans, design drainage, waste, and vent systems, meet code requirements, and make an accurate take-off for materials and labor. Includes sample cost sheets, manhour production tables, complete illustrations, and all the practical information you need to accurately estimate plumbing costs. **224 pages, 8½ x 11, $17.25**

Estimating Electrical Construction

A practical approach to estimating materials and labor for residential and commercial electrical construction. Written by the A.S.P.E. National Estimator of the Year, it explains how to use labor units, the plan take-off and the bid summary to establish an accurate estimate. Covers dealing with suppliers, pricing sheets, and how to modify labor units. Provides extensive labor unit tables, and blank forms for use in estimating your next electrical job. **272 pages, 8½ x 11, $19.00**

Carpentry Estimating

Simple, clear instructions show you how to take off quantities and figure costs for all rough and finish carpentry. Shows how much overhead and profit to include, how to convert piece prices to MBF prices or linear foot prices, and how to use the tables included to quickly estimate manhours. All carpentry is covered: floor joists, exterior and interior walls and finishes, ceiling joists and rafters, stairs, trim, windows, doors, and much more. Includes sample forms, checklists, and the author's factor worksheets to save you time and help prevent errors. **320 pages, 8½ x 11, $25.50**

Concrete Construction & Estimating

Explains how to estimate the quantity of labor and materials needed, plan the job, erect fiberglass, steel, or prefabricated forms, install shores and scaffolding, handle the concrete into place, set joints, finish and cure the concrete. Every builder who works with concrete should have the reference data, cost estimates, and examples in this practical reference. **571 pages, 5½ x 8½, $20.50**

Builder's Guide to Accounting Revised

Step-by-step, easy to follow guidelines for setting up and maintaining an efficient record keeping system for your building business. Not a book of theory, this practical, newly-revised guide to all accounting methods shows how to meet state and federal accounting requirements, including new depreciation rules, and explains what the tax reform act of 1986 can mean to your business. Full of charts, diagrams, blank forms, simple directions and examples. **304 pages, 8½ x 11, $17.25**

Builder's Office Manual Revised

Explains how to create routine ways of doing all the things that must be done in every construction office — in the minimum time, at the lowest cost, and with the least supervision possible: Organizing the office space, establishing effective procedures and forms, setting priorities and goals, finding and keeping an effective staff, getting the most from your record-keeping system (whether manual or computerized). Loaded with practical tips, charts and sample forms for your use. **192 pages, 8½ x 11, $15.50**

Painter's Handbook

Loaded with "how-to" information you'll use every day to get professional results on any job: The best way to prepare a surface for painting or repainting. Selecting and using the right materials and tools (including airless spray). Tips for repainting kitchens, bathrooms, cabinets, eaves and porches. How to match and blend colors. Why coatings fail and what to do about it. Thirty profitable specialties that could be your gravy train in the painting business. Every professional painter needs this practical handbook. **320 pages, 8½ x 11, $21.25**

Craftsman

Craftsman Book Company
6058 Corte del Cedro
Carlsbad, CA 92009

10 Day Money Back GUARANTEE

In a hurry?

We accept phone orders charged to your MasterCard or Visa. Call (619) 438-7828

Name (Please print clearly)

Company

Address

City/State/Zip

Total Enclosed _____
(In California add 6% tax)

Use your ☐ Visa ☐ MasterCard

Card # _____

Exp. date _____ Initials _____

☐ 17.50 Basic Plumbing with Illust.
☐ 30.00 Berger Building Cost File
☐ 11.25 Blprt Read. for Blding Trades
☐ 17.25 Builder's Guide to Acctg. Rev.
☐ 11.00 Bldr's Guide to Const. Fin.
☐ 15.50 Builder's Office Manual Revised
☐ 14.00 Building Cost Manual
☐ 11.75 Building Layout
☐ 25.50 Carpentry Estimating
☐ 19.75 Carp. for Residential Const.
☐ 19.00 Carp. in Commercial Const.
☐ 16.25 Carpentry Layout
☐ 17.75 Comp.: The Blder's New Tool
☐ 10.00 Concrete and Formwork
☐ 20.50 Concrete Const. & Estimating
☐ 20.00 Const. Estimating Ref. Data
☐ 22.00 Construction Superintending
☐ 19.25 Const. Surveying & Layout
☐ 16.25 Cont. Guide to the Blding Code
☐ 16.75 Contractor's Survival Manual
☐ 16.50 Cont.Year-Round Tax Guide
☐ 15.75 Cost Rec. for Const. Est.
☐ 9.50 Dial-A-Length Rafterrule
☐ 18.25 Drywall Contracting
☐ 13.75 Electrical Blueprint Reading
☐ 25.00 Electrical Const. Estimator

☐ 19.00 Estimating Electrical Const.
☐ 17.00 Estimating Home Blding Costs
☐ 17.25 Estimating Plumbing Costs
☐ 21.50 Esti. Tables for Home Building
☐ 22.75 Exca. & Grading Handbook, Rev.
☐ 9.25 E-Z Square
☐ 10.50 Finish Carpentry
☐ 21.75 Hdbk of Const. Cont. Vol. 1
☐ 24.75 Hdbk of Const. Cont. Vol. 2
☐ 14.75 Hdbk of Modern Elect. Wiring
☐ 15.00 Home Wiring: Imp., Ext., Repairs
☐ 17.50 How to Sell Remodeling
☐ 24.50 HVAC Contracting
☐ 20.25 Manual of Elect. Contracting
☐ 18.75 Manual of Prof. Remodeling
☐ 13.50 Masonry & Concrete Const.
☐ 18.50 National Const. Est.
☐ 23.75 Op. the Tractor-Loader-Backhoe
☐ 23.50 Pipe & Excavation Contracting
☐ 19.25 Paint Contractor's Manual
☐ 21.25 Painter's Handbook
☐ 13.00 Plan. and Design. Plumbing Sys.
☐ 21.00 Plumber's Exam Prep. Guide
☐ 18.00 Plumber's Handbook Revised
☐ 18.25 Proc. & Ind. Pipe Estimating
☐ 14.25 Rafter Length Manual

☐ 18.50 Remodeler's Handbook
☐ 11.50 Residential Electrical Design
☐ 18.25 Residential Wiring
☐ 22.00 Roof Framing
☐ 9.25 Roofers Handbook
☐ 16.00 Rough Carpentry
☐ 24.00 Spec Builder's Guide
☐ 13.75 Stair Builder's Handbook
☐ 15.50 Video: Asphalt Shingle Roofing
☐ 15.50 Video: Bathroom Tile
☐ 15.50 Video: Drywall Installation
☐ 15.50 Video: Electrical Wiring
☐ 15.50 Video: Exterior Painting
☐ 15.50 Video: Finish Carpentry
☐ 15.50 Video: Hanging An Exterior Door
☐ 15.50 Video: Int. Paint & Wallpaper
☐ 15.50 Video: Kitchen Renovation
☐ 15.50 Video: Plumbing
☐ 80.00 Video: Roof Framing 1
☐ 80.00 Video: Roof Framing 2
☐ 15.50 Video: Rough Carpentry
☐ 15.50 Video: Windows & Doors
☐ 15.50 Video: Wood Siding
☐ 7.50 Visual Stairule
☐ 11.25 Wood-Frame House Const.

In a Hurry? Call (619) 438-7828 Orders taken on Visa or MasterCard   crce card

Craftsman

Craftsman Book Company
6058 Corte del Cedro
Carlsbad, CA 92009

10 Day Money Back GUARANTEE

In a hurry?

We accept phone orders charged to your MasterCard or Visa. Call (619) 438-7828

Name (Please print clearly)

Company

Address

City/State/Zip

Total Enclosed _____
(In California add 6% tax)

Use your ☐ Visa ☐ MasterCard

Card # _____

Exp. date _____ Initials _____

☐ 17.50 Basic Plumbing with Illust.
☐ 30.00 Berger Building Cost File
☐ 11.25 Blprt Read. for Blding Trades
☐ 17.25 Builder's Guide to Acctg. Rev.
☐ 11.00 Bldr's Guide to Const. Fin.
☐ 15.50 Builder's Office Manual Revised
☐ 14.00 Building Cost Manual
☐ 11.75 Building Layout
☐ 25.50 Carpentry Estimating
☐ 19.75 Carp. for Residential Const.
☐ 19.00 Carp. in Commercial Const.
☐ 16.25 Carpentry Layout
☐ 17.75 Comp.: The Blder's New Tool
☐ 10.00 Concrete and Formwork
☐ 20.50 Concrete Const. & Estimating
☐ 20.00 Const. Estimating Ref. Data
☐ 22.00 Construction Superintending
☐ 19.25 Const. Surveying & Layout
☐ 16.25 Cont. Guide to the Blding Code
☐ 16.75 Contractor's Survival Manual
☐ 16.50 Cont.Year-Round Tax Guide
☐ 15.75 Cost Rec. for Const. Est.
☐ 9.50 Dial-A-Length Rafterrule
☐ 18.25 Drywall Contracting
☐ 13.75 Electrical Blueprint Reading
☐ 25.00 Electrical Const. Estimator

☐ 19.00 Estimating Electrical Const.
☐ 17.00 Estimating Home Blding Costs
☐ 17.25 Estimating Plumbing Costs
☐ 21.50 Esti. Tables for Home Building
☐ 22.75 Exca. & Grading Handbook, Rev.
☐ 9.25 E-Z Square
☐ 10.50 Finish Carpentry
☐ 21.75 Hdbk of Const. Cont. Vol. 1
☐ 24.75 Hdbk of Const. Cont. Vol. 2
☐ 14.75 Hdbk of Modern Elect. Wiring
☐ 15.00 Home Wiring: Imp., Ext., Repairs
☐ 17.50 How to Sell Remodeling
☐ 24.50 HVAC Contracting
☐ 20.25 Manual of Elect. Contracting
☐ 18.75 Manual of Prof. Remodeling
☐ 13.50 Masonry & Concrete Const.
☐ 18.50 National Const. Est.
☐ 23.75 Op. the Tractor-Loader-Backhoe
☐ 23.50 Pipe & Excavation Contracting
☐ 19.25 Paint Contractor's Manual
☐ 21.25 Painter's Handbook
☐ 13.00 Plan. and Design. Plumbing Sys.
☐ 21.00 Plumber's Exam Prep. Guide
☐ 18.00 Plumber's Handbook Revised
☐ 18.25 Proc. & Ind. Pipe Estimating
☐ 14.25 Rafter Length Manual

☐ 18.50 Remodeler's Handbook
☐ 11.50 Residential Electrical Design
☐ 18.25 Residential Wiring
☐ 22.00 Roof Framing
☐ 9.25 Roofers Handbook
☐ 16.00 Rough Carpentry
☐ 24.00 Spec Builder's Guide
☐ 13.75 Stair Builder's Handbook
☐ 15.50 Video: Asphalt Shingle Roofing
☐ 15.50 Video: Bathroom Tile
☐ 15.50 Video: Drywall Installation
☐ 15.50 Video: Electrical Wiring
☐ 15.50 Video: Exterior Painting
☐ 15.50 Video: Finish Carpentry
☐ 15.50 Video: Hanging An Exterior Door
☐ 15.50 Video: Int. Paint & Wallpaper
☐ 15.50 Video: Kitchen Renovation
☐ 15.50 Video: Plumbing
☐ 80.00 Video: Roof Framing 1
☐ 80.00 Video: Roof Framing 2
☐ 15.50 Video: Rough Carpentry
☐ 15.50 Video: Windows & Doors
☐ 15.50 Video: Wood Siding
☐ 7.50 Visual Stairule
☐ 11.25 Wood-Frame House Const.

Craftsman Book Company, 6058 Corte Del Cedro, P. O. Box 6500, Carlsbad, CA 92008   crce card

BUSINESS REPLY MAIL
FIRST CLASS   PERMIT NO. 271   CARLSBAD. CA

POSTAGE WILL BE PAID BY ADDRESSEE

Craftsman Book Company
6058 Corte Del Cedro
P. O. Box 6500
Carlsbad, CA 92008—9974

NO POSTAGE
NECESSARY
IF MAILED
IN THE
UNITED STATES

BUSINESS REPLY MAIL
FIRST CLASS   PERMIT NO. 271   CARLSBAD. CA

POSTAGE WILL BE PAID BY ADDRESSEE

Craftsman Book Company
6058 Corte Del Cedro
P. O. Box 6500
Carlsbad, CA 92008—9974

NO POSTAGE
NECESSARY
IF MAILED
IN THE
UNITED STATES

BUSINESS REPLY MAIL
FIRST CLASS   PERMIT NO. 271   CARLSBAD. CA

POSTAGE WILL BE PAID BY ADDRESSEE

Craftsman Book Company
6058 Corte Del Cedro
P. O. Box 6500
Carlsbad, CA 92008—9974